Mayne Reid

The naturalist in Siluria

(Herefordshire, Radnorshire, Brecknockshire and Glamorganshire)

Mayne Reid

The naturalist in Siluria
(Herefordshire, Radnorshire, Brecknockshire and Glamorganshire)

ISBN/EAN: 9783744715904

Printed in Europe, USA, Canada, Australia, Japan

Cover: Foto ©berggeist007 / pixelio.de

More available books at **www.hansebooks.com**

THE

NATURALIST IN SILURIA.

FRONTISPIECE.

THE
NATURALIST IN SILURIA.

(HEREFORDSHIRE, RADNORSHIRE, BRECKNOCKSHIRE AND GLAMORGANSHIRE).

BY

CAPTAIN MAYNE REID,

Author of "The Scalp Hunters," "The Death Shot," etc.

THE MARTIN.

PHILADELPHIA:

GEBBIE & CO.

1890.

PUBLISHERS' PREFACE
TO THE NATURALIST IN SILURIA.

THIS new book is a departure from the class of literature by which Captain Reid made his name famous. It was left in MS. at his death in 1883, and not until 1889 did his widow place it in the hands of a publisher.

Any careful reader of Captain Mayne Reid's Tales of Adventures will have discovered that his strongest point is his vigorous and accurate description of Natural Scenery: Therefore, when he retired from his travels and London Society to Penyard Grange, in the lovely valley of Woolhope, not far from the city of Hereford, it was only a matter of course that the old Indian hunter should take to the woods and fields, and devote his leisure to the study of Natural History—examples for which existed so abundantly in his immediate neighborhood.

A more ardent lover of nature than our author it would be hard to name—not even excepting the celebrated Gilbert White, of Selbourne.

The district in which he made his observations is indicated in the title of his book, *Siluria*, concerning which we quote from *Camden's Britannia:* "Ptolemy says the Silures inhabited those countries which the Welsh call by one general name, *Dehubarth*, or the southern part; branched at this day into new names—Herefordshire, Radnorshire, Brecknockshire and Glamorganshire."

Professor Murchison revived the forgotten name of the tribe of ancient Britons by naming the palæozoic rocks discovered in that locality Silurian, a term which has been adopted by all geologists in describing this particular strata—no matter where discovered.

GEBBIE & CO.

LIST OF ILLUSTRATIONS.

INTRODUCTORY.

A NATURALIST'S PARADISE.

I DWELL in a district of country remarkable for its richness in plant and animal life ; I mean, of course, the wild
and indigenous. So varied and plentiful are the species
that in these respects I venture to believe there is no
other part of England, or, indeed, the United Kingdom,
which can at all compare with it. This profusion is
chiefly due to its peculiar geological features. As will be
easily understood, the geology of any particular part of
the earth's surface affects the character of its botany so
much that the former may appropriately be termed the
parent of the latter ; while, in turn, the plant-life may be
regarded as the creator and nursing-mother of all that
" lives, moves, and hath being." If, for instance, some
grand upheaval—volcanic, plutonic, or by whatever name
called—have tossed to the surface a varied series of the
stratified rocks which form the earth's crust, and left
their tilted edges exposed to the atmosphere, there will
spring up on them a varied veg tation, with animal life
in like manner diversified. And it will also be obvious
that the more abrupt the dip of the upheaved strata the
greater will be this variety within the limits of a given
district ; as, of course, the sharper the angle of elevation
the narrower the exposed surface of any particular stratum.

Now I am living in the immediate neighbourhood of

more than one such upheaval; but one so remarkable as to have a world-wide repute. For my residence is in *Siluria*, contiguous to that singular and symmetrical "valley of elevation" known as Woolhope. From the summit of a high wooded hill, Penyard, which rises abruptly in rear of my house, I can look over the whole series of Upper Silurian rocks, from the northern edge of their upcast at Mordiford, near the city of Hereford, to their southern projection by Gorstley in Gloucestershire. There they dip under the Devonian or Old Red Sandstone, again to show upon the surface a little farther south, in the smooth-rounded dome of May Hill, standing solitary, with its crest of Scotch firs conspicuous from afar.

Looking to the right or east of the Woolhope district, though still northward from my point of view on Penyard, I have before my eyes, and at less than fifteen miles of direct distance, the bold isolated chain of the Malverns, an elevation geologically remarkable as that of Woolhope itself. For while in its central axis we have all the metamorphic rocks—schists, both micaceous and hornblende, with granite, syenite, gneiss, and felstone—as the Laurentian, the oldest sedimentary formation known— there also is the product of Plutonic action in Trappean rocks, basalt brought to the surface in shafts and dykes— volcanic too, the Raggedstone Hill at the southern extremity of the range being itself an ancient crater. Again, on its western flank are the Silurian strata exposed by upheaval, and the denudation of the Old Red; at the same time that a corresponding downfall along its eastern base—a fault of possibly many miles in vertical measurement—shows us the more recent Triassic formation overspreading the beautiful plain or "vale" of Worcester, with a little farther off the overlying Lias, here and there rising into isolated hills, capped by the yet more recent Oolite.

Still nearer, however, to my point of observation are these secondary deposits, for their western edge approaches the Palæozoic rocks not far from the foot of May Hill— from me little more than a league off.

Westward, and in fact all round me, extends the Old Red Sandstone, the characteristic rock of Herefordshire, as also tho adjacent county of Monmouth. Its strata of 10,000 feet thickness—variously composed, and not all of a red colour, as might be supposed by the misleading of a name—in many places give evidence of the most violent convulsion, their *dip* observable at angles of every degree. Beyond doubt, throughout Herefordshire and Monmouth the Old Red was once overlain by rocks of more recent formation ; certainly by the Carboniferous, whose seams still cover it in the South Wales coalfield, the Clee Hills, and Forest of Dean. Than this, to the geologist, there is no more interesting district in England—I might say in all the world. For within a remarkably limited circle the view on one side embraces the whole of the upper and lower Palæozoic rocks, with all the Mezozoic, excepting the Cretaceous ; and on the other the Trias, Lias, and Oolite ; while near by, on the west, lies the valley of the Wye, rich in drifts of geological interest, and eastward the wider and more extended valley of the Severn, itself an ancient sea-bed.

Turning southward, I have the Forest of Dean before my face, a tract of country singular as it is celebrated. It is, in point of fact, an elevated table-land, several hundred feet above the level of the plains around, here and there intersected by deep ravines, but on all sides presenting a façade, steep, almost precipitous. My dwelling is contiguous to its northern edge,—Penyard Hill being but an out-lier of it,—and, though my house and grounds are on the Old Red, a cannon fired from the front door, with sufficient elevation of aim, would fling its shot over the wooded brow of the Forest, into the "lower coal measures." But before it rolled to rest among these, the ball, obliquing upwards, would first pass over a bed of Red Conglomerate, mixed with marl and other sandstones; next cutting across a belt of yellow sands with red marls, and sands of this same colour; then a tract of Mountain Shale and Mountain (carboniferous) Limestone; after this, a stratum of Millstone Grit, and another of Upper Sandstone, with seams of clay and marls; crossing the crest of this elevated plateau, and passing on finally to fall among the above-mentioned "coal measures"; which, quoting the words of an eminent geologist, "are a relic of the most profuse vegetation the world has ever beheld."

It may seem strange that a section of country so signalized in the countless ages past should still possess a character in correspondence. But it is even so, its *flora* being abundant beyond any other I know of. Within a circle of 20 miles radius around my house, I find between 600 and 700 species of phanerogamous plants, while the cryptogamia are alike plentiful. If the theory advanced be admitted, it would follow that the *fauna* is proportionately rich; and so, in reality, it is. As proof sufficient— and, to me, rather more than satisfactory—the fox and badger prey upon my poultry, assisted in their depredations by the pole-cat, weasel, and stoat; while hares and rabbits crop the cabbages in my kitchen-garden. The otter bathes its sleek body in a brook—an influent of the Wye—which meanders through my ornamental grounds;

the water-vole (*Arvicola amphibia*) plunges in my fish-pond, and honeycombs the banks of the self-same brook that supplies it; while its congener of the land (*A. agrestis*) breeds in myriads over the adjoining meadows, hollowing out its nest just enough under the sward for its hairless callow young to be clear of the dangerous scythe-blade. Around the drier ditches the hedgehog searches for snails, munching these crustacea, despite their silicious shield—which is no protection against the teeth of the urchin, who swallows armour and all. The mole, "mooting" after earth worms, if not kept under by continuous trapping, makes spoil of my pasture-land, in places giving it the appearance of a ploughed field; while the squirrel, more agile, and less destructive, lends animation to my groves and copses. Not so nice is the near companionship of the rat,—he erroneously supposed to be a native of Norway,—who ranges around my rick-yard, occasionally seeking entrance into barn and corn-bin, with a suspicion attached to him of not being content with a *menu* purely vegetarian, but having also a tooth for young chicks and ducklings. When I add to this list of indigenous mammals the mouse, dormouse, and several species of *Sorex*, the catalogue is pretty complete; though I have a *soupçon* of a wild cat, which seems to have shown itself in the neighbourhood some months ago. I am in search of this suspicious "Tom," and if I can "tree" him will account it a triumph.

The reptile world around me is represented by the usual British genera and species: two snakes and a doubtful third, the "slow-worm," sometimes called "blind-worm" (*Anguis fragilis*), of which last I have lately captured a specimen measuring eighteen inches in length. Batrachians abound in the shape of toads, frogs,

and newts, these last hideous creatures of at least two
distinct species, the common or " smooth " (*Triton punc-
tatus*), an alligator in miniature, and the crested, or
" warty " (*T. cristatus*), which more resembles the real
Nilotic crocodile, or its congener of the Senegal. Lizards,
too, are among our land reptiles, and with these most
modern naturalists class the slow-worm, which, though
having some affinities with the lizard family, is certainly
nearer to the serpent in habit, as it is in "personal"
appearance.

Of fishes we have the usual freshwater species; my
brook and pond, however, yielding only trout, eels, minnows, and the wicked little bullhead (*Cottus gobio*), which,
incautiously taken up in the hand, bites like a very shark.
But below, in the " wandering Wye," the salmon, king
of fish, holds court, having for his subjects a variety of
finny and scaly creatures, among them the famed lam-
prey, a delicate morsel, though it *did* prove indigestible
in the stomach of a king.

Insects ? Ah! we have them in swarms, myriads;
the Wye's valley being a very garden of Eden for the
entomologist, who may here fling his net over butterflies
bright as summer flowers, and capture Scarabæans of
hues vivid as tropic sun ever shone upon. But he had
well beware when seeking them; for by the dry banks
" whereon the wild thyme grows " are wasps and hornets;
and amid the lush vegetation of the moist Wyeside woods
sings and stings the mosquito (*Culex pipiens*), while the
" harvest bug " (*Leptus autumnalis*), a near relative of
the West Indian "jigger," if not the veritable thing itself, though no larger than a particle of Cayenne pepper,
which it ludicrously resembles, inserts its tiny nippers
into the skin—the result an intolerable itching.

As the good host who reserved his choicest " bin " to the close of the feast, so I have kept back my best and greatest favourites—the birds. Of these, the now too rare and stately heron cranes its long neck, and projects its bayonet-like beak over the afore-mentioned fish-ponds on the look-out for a speckled trout, or, it may be, a slippery eel; while the kingfisher darts past like an arrow, showing its back of turquoise blue, the food of its selection being the smaller fry of minnows and bull-heads.

In the same water the pretty moor-hen disports herself, and with coquettish strut makes frequent promenades upon my lawn, fearlessly coming on over the carriage-sweep, and up to the steps of the door-porch. Nor has she the smooth turf all to herself, for the ring-dove, or cushat (*Columba palumbus*) also alights upon it, to look after beech-mast and acorns, occasionally accompanied by its near congener, the stock-dove (*C. œnas*); while the more slender turtle (*C. turtur*) flies past, keeping farther a-field. All three have their nests near, and their cooing sounds pleasant to my ears, telling me aught but a " sorrowful tale."

On the same verdant sward the noisy jay shows itself, coming so close to the drawing-room windows that an artist seated in one of them might take the portrait of this beautiful bird; not with the dim, damaged lustre of a stuffed specimen or caged captive, but in all the radiant hues of life, liberty, and action. Quite as often the green woodpecker (*Picus viridis*)—a bird of such brilliant plumage as to have obtained the title of " English parrot "—drops down upon the lawn, to do me an essential service by delving its long beak into the ant-hills which infest the sward, and destroying thousands of

these too-industrious insects. In the woods I occasionally hear the tap-tapping of his two cousins, the great
and little-spotted woodpeckers (*P. major* and *P. minor*);
but these are much rarer in the neighbourhood. Not so
the magpie, here only too plentiful. He hops about
among the tall fir-trees where he has nested, or makes
descent upon the grass pastures, at intervals alighting on
the lawn to pick up some morsel that may there have
caught his eye. But the cunning chatterer remains only
a moment; for he has been guilty of " fowl " play in the
poultry-yard, and, knowing it, dreads my gun. Out in
the fields the carrion crow (*Corvus corone*), also a foe to
the chirping chicks, stalks majestically, grubbing among
mole-heaps and the deposits of animal ordure, also gobbling up field-mice; while his cousins-german, the rook
and jackdaw, more satisfied with a vegetable diet, seek
it everywhere over the ploughed and pasture lands, in
concert, and consorting with, clouds of starlings.

Of birds more properly called predatory there is no
scarcity here. The sparrow-hawk courses low along the
hedges; while the kestrel, of bolder flight, hovers aloft,
as if suspended on an invisible string, at intervals changing his point of aëriel observation, to hover again, or
swoop down upon the prey he has marked for a meal.
The buzzard (*Falco buteo*) is not unfrequently seen soaring
over Penyard's wooded hill, and also the peregrine falcon
(*F. peregrinus*), while the great kite (*F. milvus*) is a less
frequent visitor. Nor are the little merlin (*F. æsalon*), the
hobby (*F. subbuteo*), the hen-harrier (*F. cyaneus*), and
even the honey buzzard (*F. apivorous*) unknown to our
neighbourhood.

The night birds of prey are here represented by the
tawny and barn owls (*Strix stridula* and *S. flammen*),

and others; while the night-jar, or goat-sucker (*Oapri-mulgus*), on equally silent wing sweeps along the wood's edge, or lights beside the sheep in pen or pasture.

With singing birds I am blessed. Summer and winter the blackbird delights me with his bold lay; the thrush making music of a more scientific strain. The lark and grey linnet also salute me throughout the diurnal hours, mingling their notes in harmony with those of the three finches—chaff, bull, and gold—all of which nest in the near trees and shrubberies. Among the humbler warblers, I can detect the twitter of several species of tits, as the blue, long-tailed, cole, and marsh; but, though not the grandest of bird melody, perhaps pleasantest to our ear is the gentle trill of the robin, for he lets us hear it throughout the chill winter-tide, when most of the more ambitious songsters are silent. In spring, however, and throughout the summer months, we have a wandering minstrel, who pays us an annual visit; and while he is with us, all our other feathered musicians, if not shamed into silence, seem, at least, to feel their inferiority. For he is primo-tenore, primo-basso, soprano, contralto—everything! Need I say that this distinguished visitor is the nightingale? Though he gives his concerts chiefly during the hours of night, and notably between midnight and morning, yet oft are we favoured with them during broad daylight. In the early part of last summer I more than once heard his matchless strain—meant, no doubt, for his mate, the "prima donna," sitting on her nest, and for the time silent—heard it in the afternoon, with a bright sun shining in the sky! Which gives contradiction to the old song,—

> " The nightingale, I've heard them say,
> Sings but at night, and not by day."

THE GEOGRAPHICAL DISTRIBUTION OF BIRDS.

The purpose of this note is to point out the difficulty of determining the boundary within which certain birds may be found, especially the smaller sorts. The most assiduous observer, ever on the watch, will now and then discover a species in his immediate neighbourhood whose existence there he had never before noted nor suspected. Yet it did exist—may even have nested and bred in his own grounds, within a hundred yards of his house. Were it not for their song or call-note betraying their presence, many of the migratory birds—our summer visitants—would come and go without the ordinary observer, and in some cases the ornithologist himself, being the wiser of it. For these are with us only when the trees are in full leaf, to screen them from curious eyes —a screen most of them know the advantage of, and take. You may hear the blackcap and garden warbler giving out their dulcet notes every day, and hour after hour, yet never get sight of either of these superb song-sters, though perched upon a spray within less than a rod's length from the spot where you are standing. But it is not alone with our summer visitants that there is this difficulty of fixing the home and habitation. It also exists, to a greater or less degree, as regards the winter ones, and even our permanently resident species, who have no tree foliage to hide behind. I speak more par-ticularly of the smaller kinds, from having lately met several instances in point: by the discovery of species in my own neighbourhood, whose existence in it I had long doubted. Yet had they been there, as I now know, their presence becoming ascertained almost by accident.

A bird of sparrow size, seen at a hundred, or even

fifty, yards distance, needs sharp eyes for its identification; and as the rarer sorts are usually the more shy, and keep farther off, all the more difficult is this, and, as a consequence, determining the *locale* of such species.

THE LOCAL DISTRIBUTION OF BIRDS.

That certain species are found in particular localities—I may say, spots—while absent from others near by, is a fact well known and seemingly singular. The reason, however, is obvious: the conditions of the places are not the same, though apparently they may be so. In one there is some sort of food—seed, berry, root, or insect—which is wanting in the other; and, as almost every species of bird has a predilection for some special diet, where this exists not neither will the bird.

But food seems not the only attraction which determines the dwelling-place of birds. Some affect the woodland shade, while others prefer the open; and still others frequent spots of an intermediate character, neither thickly overgrown with trees, nor yet altogether treeless. Dryness, moistness, and water—stagnant or running—are also influencing factors; and so too the configuration of the ground, whether it be hilly or level, the altitude of the hills, and the exposure of their slopes in relation to the points of the compass. Certainly food is not the only thing which influences birds in their choice of habitat, as we have evidence in the preference shown by the common house-sparrow. A pair, or at most two pairs, of these noisy chatterers haunt around my house, and breed by it; while at every farmstead in the neighbourhood a large flock may be seen at all times, both in summer and winter. Yet there is a farmyard attached to my establishment, with plenty of pickings for the *fringillidæ*. No doubt the reason for the sparrows keeping away from my premises is because the house, outbuildings and all, is overshadowed by tall trees, and the *passer domesticus* prefers to perch on hawthorn bush or bare gable-end.

VEGETATION ON THE OLD RED.

The soil of the Old Red Sandstone seems wonderfully congenial to certain plants of the order *compositæ*. At least some strata of it are so, for in a system of rocks 10,000 feet thick, and deposited during countless ages, there must be much variety in the nature of the deposited substances.

I here speak of strata high up in the system, close to the Carboniferous, but under the shales and Conglomerate of the Old Red itself. In my kitchen garden, whose soil is over a seam of this kind, there grow Jerusalem artichokes that remind me of the tropics, recalling a brake of bamboo cane. A six-foot man standing on the back of a sixteen-hands horse could not touch their tops with his hand upraised to its highest; an average stalk, which I have submitted to measurement, proving to be 13 feet 3 inches—without reckoning the roots—and having a girth of 4¾ inches! Not a bad growth for temperate zone vegetation, within a period of less than six months.

I believe that both the Jerusalem artichoke and its near congener, the sunflower (*Helianthus annuus*), might be profitably cultivated in this district; the former not only for its tubers, but the stalks and leaves as an article of fodder; while the seeds of the latter are well known to be nourishing food for poultry, fowls and turkeys being alike fond of it.

In an orchard adjoining this garden, up against a dry bank at the back, I some time ago observed a thistle of such extraordinary dimensions that I placed myself alongside its stalk, to make a rough estimate of its height. The crown of my hat did not reach half-way to the top, nor anything like it; while its stem by the base was nearly as thick as my wrist. It was one of the sort which are here commonly called "boar thistles"; but I took no exact note of the species, determined on having it home and submitting it to rule and tape. As ill luck would have it, the discovery of this vegetable giant was made on a Sunday, which caused the deferring of my intention to the following day. Then revisiting the spot, with my gardener and a grubbing tool, I had the mortification to find it gone. A right-of-way path runs by, near the bank where it grew, and some villanous trespasser, whom I cannot help being angry at, had taken a fancy to this gigantic *carduus*, torn it up by the roots, and carried it clean away. As there is no Scotchman dwelling in the immediate neighbourhood, I am puzzled about the motive of the pilferer. It may have been botanical curiosity, or only an idle freak; though I have heard that the birdcatchers sometimes use these large thistles, *limed*, for the taking of goldfinches—an explanation of the rape probable enough. It was certainly as tall as any of my artichokes, and the stalk near the base of much greater thickness.

Besides the composite, other plants grow luxuriantly on the Old Red. In the same garden carrots and parsnips attain the dimensions of a man's forearm; while beet-root needs sowing late to keep it within bounds for table use. Some of my "Mexican" potatoes, planted in it this year, threw up haulms so rank and high I had the curiosity to measure one. It was over seven feet in length, exclusive of tubers and rootlets, which would have made it at least six inches more.

This same year, in the aforementioned orchard adjoining—of three acres area—was reaped a crop of oats that threshed out over 200 bushels, or seventy to the acre; this in damp soil, and under the shadow of six score apple and pear trees, all old and umbrageous, and despite the culms having been broken and "laid" by heavy rainstorms long ere the corn could ripen. When green and standing erect, they look like a sedge of bulrush. A boy sent in among them to search for a landrail's nest was buried above the head, and soon lost to my sight. Drawing one, I laid it alongside the measuring-rule, to find it 5 feet 3 inches in length, and thicker than the quill of a swan. So much for the fertility of the soil over the Old Red.

From all which it may be deduced that a farmer designing to take a new farm, or make the best use of his old one, should know something of geology.

HOW THE YEARS VARY.

No one who keeps a diary of Nature's doings can fail being struck with how they are diversified in the different years. The variation is, of course, chiefly due to atmospheric influences, but the operation of these is the question difficult to answer. As who can say why one winter is of the mildest, snowless, almost without frost; whilst another is of rigorous severity? Phenomena so marked are noticed by all; but the naturalist alone takes note of their effect on the world of living organisms, vegetable and animal; on dead matter, too, for that is also influenced by them.

As illustrating this diversity, let us take a particular district of country, as, for instance, that in which I myself dwell. Being upon the Old Red Sandstone, it is much affected by ants of several species; so much that they are accounted a pest, the yellow ant (*Formica flava*) certainly being this. In the summers of 1878-79-80 these insects swarmed upon the pastures, throwing up their "tumps," deleterious to the growth of nutritive grasses; while during the summer of 81 only a few were observable. This seems all the more strange from the previous summer being dry and warm, as one would suppose favourable to ant life, while those preceding were the reverse. I have a somewhat similar record of the common house fly (*Musca carnaria*), whose scarcity for several years past has been notable. But though appearing early in the spring, in summer it seemed to have ceased existence, while a species much resembling, and commonly mistaken for it, the biting *Stomoxys calcitrans*, was unusually abundant.

Again, wasps, that in several previous years did much damage in our pear orchards, and were even a source of annoyance to mowers and reapers, in the autumn of 1881 were little seen or heard of. In the spring also "blight," caused by grubs of countless kinds, so abounded that many trees—notably oaks—were entirely stripped of their foliage, and stood with branches bare as in winter, till the flowing of the midsummer sap gave them a fresh livery of leaves. For years before there had been little or nothing of this larval devastation.

Going underground to the earthworm (*Lumbricus terrestris*), I noticed that for several years past my lawn was remarkably clear of their castings, yet in the autumn of

1881 they reappeared thickly over the sward, and ever since there has been a nightly renewal of them.

In the ornithological world these alternations are equally observable. The blackbird and song-thrush lead an undisturbed life in my grounds, where for years past, during their season of song, there was no day, scarce even an hour, without the strain of one or the other being heard. But, strange to say, throughout the spring and summer of 1881 it was something unusual to hear the note of either merle or mavis; all the stranger, from the fact of both birds seeming to be about in even more than their usual numbers.

The yellow-hammer is known to be a gregarious bird; but, so far as I have observed, oftener consorting with other species than exclusively with its own kind; even then being in the minority, its lemon-coloured crest and breast appearing in an assemblage of other *fringillidæ* but as one to five or six. In the autumn of 1881, how-ever, and up till now, I have frequently observed flocks of yellow-hammers, numbering two to three dozen indi-viduals, in various places, and quite apart from birds of other species, their abundance seeming to me as strange as this segregation.

If we turn to the starlings, we find a like variation at different periods of time. During the breeding season of the years 1878-79-80, after the eggs had been hatched, a glance cast skyward above my house and grounds was almost sure of being rewarded by the sight of a starling on return to its nest with a grub in its beak, or taking departure therefrom in quest of another. Yet in 1881 these journeyings to and fro were so seldom witnessed that it seemed as if this, one of our commonest birds, had become a *rara avis !*

And, as most people know, the migratory birds are more plenteous in one year than another, notably the nightingales in summer, and the fieldfares and redwings in winter. Their presence or absence, however, can be accounted for by the temperature, with other climatic changes; and, no doubt, these are the chief influencing causes throughout all, though we are ignorant of the *modus operandi.*

OUR WILD PIGEONS.

Of all our native birds, none seem to me invested with more vivid interest than the wild pigeons. I cannot help regarding them as the greatest ornament and truest emblem of sylvan scenery; and I never see one sitting upon a bough, or in bold, earnest flight through the air, without a sense of exquisite pleasure—a feeling of thankfulness that my home is in the country.

RING AND ROCK DOVES.

In addition to the physical beauty of these birds, their MORAL character—if I be permitted so to speak—is unexceptionable. They are neither predatory nor polygamous; for the first, contenting themselves with a simple vegetarian diet, while in their marital relations they are models of constancy and affection. No lover sues to his sweetheart in gentler or more pleading tone; and he were a good husband who will show half the attention to his wife which the dove does to its mate.

Having made use of the word *dove*, I may here remark that all pigeons were formerly known as doves, even the domestic variety being so called. Hence the origin of the name " dovecote," the thing itself being in reality a pigeon-house, which in past times was an appanage of every mansion and monastery in the land; one of such importance, too, that statutes were enacted limiting their number, even to the right of having them at all—just as our Game Laws of the present day.

It was not for mere ornament or fancy, dovecotes were kept, but with a view to the more substantial benefit derived from them in supplying a choice article of food. They took rank with the fish-pond; beside which they often stood, at a time when the fishmonger and poulterer had either no existence or dwelt at an inconvenient distance.

When the name *pigeon*—an Anglicized form of the Italian *pigione*—came into general use in this country is not very clear, though now it is universally employed when speaking of the larger species of the genus *columba*, while the original designation of dove is still retained for the smaller ones. Audubon has ventured on a distinction, giving the name *pigeon* to those that make their nests in large numbers on the same tree; while the *dove* is solitary in its nidification. The American naturalist, not always accurate, was evidently misled by the habits of the species which came under his observation—a very limited number. The facts are all against his speculative theory, most of the European species nesting apart, and only gregarious when in search of food and the breeding season is over.

I believe it is not generally understood, though of course known to naturalists, that in England we have four distinct species of the genus *Columba*, three of which are called pigeons, the fourth a dove. This is exclusive of the tame or House pigeon, and its numerous varieties.

Of the wild sorts, the first in point of size, as the most commonly distributed, is the *Queest* or *Quest*, also called *Cushat*. It is the Ring-dove (*Columba palumbus*) of the ornithologists, a name supposed to have originated in the whitish blotch on both sides of its neck, bearing resemblance to a ring. The similitude is very slight, and the title altogether inappropriate. More correct is its common appellation "Quest," evidently a derivation from the Latin *quæstus*, since it is of this species Henry Kirk White speaks as oft warbling "its sorrowful tale."

A noble bird it is, in symmetry of form far excelling any of the domesticated breeds, while in size it is also their superior. And as an article for the table, it not only excels the tame sorts, but if put into the hands of a cook who knows his or her business, in point of *goût* it equals the very best of our game birds. Give it the same treatment as a grouse, and if your palate be not regulated by fashion or caprice, you will never after care to pay 7*s.* the brace for moor-cocks while you can buy cushats at 2*s.* the pair. As for partridges, why they should sell, pound for pound, at three times as much as wild pigeons, is a question I leave to the game dealers. But one which puzzles me yet more is, that a House pigeon, also pound for pound, commands nearly double the price of its wood-dwelling congener. As a thing to be eaten, there is no comparison between the two, the wild being as much superior to the tame as a pheasant to a barn-door fowl.

The quantity of food supply derived from this source is deserving of serious consideration. It is difficult, perhaps impossible, to estimate the exact amount; but from the numbers of these birds brought into the market, and the hundreds of thousands besides that go to the table without ever having appeared on a poulterer's stall, some idea may be deduced of their commercial value. And it is worth remembering, that in this case the cost of production is altogether disproportioned to the value produced, compared with that of barn-door fowls, or even pigeons of the domesticated kind. Tho Quest may do a little damage at seed-time and among the summer tares and peas, or, in a very severe winter, peck holes in the turnips, and eat off their tops,—but the House pigeon has to be credited with the same.

It has been said that the turnip diet renders the flesh of the wild species so rancid as to be unpalatable. The naturalist of Selborne first made this assertion, and it has been repeated by other writers over and over again. I have not found it correct; and during times of frost and snow I have had an excellent opportunity of testing its accuracy. Never was there better, for, by the complete failure of our usual berry crop, the wild pigeons have then had no other provender than turnips; and although I have eaten several that were shot in the very act of feeding on these vegetables, I could perceive nothing of the rancidity spoken of.

That the Quest is not the progenitor of our domestic birds has been generally admitted. The very different modes of their nidification is, to a certain extent, proof

of this : the one nesting in trees, the other never. Three
years ago I would have added that only the one *perched*
upon trees, the other never. But I have of late had evidence that this, though in accordance with the universal
belief, would be incorrect.

In my stable-yard there is a proper pigeon-house,
which for some time had been untenanted. Three years
ago I re-stocked it with some half-dozen pairs, among
which there were most of the sports or varieties of Fan-
tails, Tumblers, Carriers, and the like. The pigeon
quarters are in a loft over the coach-house, the entrance
to them being through a network of holes in the gable,
close to which grow several tall trees, beeches, limes,
chestnuts, and oaks. Some days after introducing the
pigeons to their new dwelling-place I was surprised to
observe them perching upon the trees; not only those
contiguous to the gable, but others full fifty yards off, in
the ornamental grounds. Nor did it seem a mere momen-
tary caprice, a dropping down upon the branches to fly
instantly up again. Instead, they sat contentedly there,
often for hours at a time. My servants, and all who saw
them thus roosted, were as astonished as myself, saying
they had never seen the like before. As several Quests
were moving about among the same trees, and occasionally
alighting upon them, I had hopes to see courtship and
marital connections established between the wild and the
tame, thus contradicting all past experience. But, no !
Beyond gazing at one another—the wild birds, no doubt,
the more astonished of the two, seeing their domain
thus intruded upon—the acquaintance went no farther.
Congeners and cousins though they were, no love, affinity,
or attachment sprang up between them.

Oddly enough, after the first few weeks the House

pigeons ceased to perch upon the trees, confining themselves to tops of walls, roofs, and chimneys; and since I
have never seen one of them set foot upon a branch.

I need not here give a description of the Quest, its
mode of nidification, nor its ordinary habits.   All this, if
not already known, can be learnt from the encyclopædias.
I will only add, that in the valley of the Wye, well
wooded everywhere, it is one of our commonest birds.
In spring and summer I could not gaze out of my window
for twenty minutes at a time without seeing one or more
sitting motionless on the branch of a tree, winging their
way through the air, or it might be walking over the
ground, constantly bowing or ducking their heads; from
which habit they derive their Latin name *columba*, from
the Greek *kolumban*, to dive.   It is also the origin of
their more correct appellation of *dove.*

The Stock-dove (*Columba œnas*) is not so common
upon the Wye, nor, I believe, anywhere in England, as
the Quest.   It is, however, anything but rare; and, although to a certain extent migratory, we have it in
Herefordshire all the year round, numbers breeding in
this neighbourhood.   It is the species which so much
puzzled the naturalist of Selborne; and, by his account,
was altogether a bird of passage in that part of the
country.   In speaking of it as not being the progenitor
of our House pigeons, he says:—" It is manifestly larger
than the common house-dove, against the usual rule of
domestication, which generally enlarges the breed."

The conclusion is not universally correct, as I can
show by a reference to the wild turkey and its tame
descendant.   But in this case even the premises are in
default, for the Stock-dove, so far from being larger
than the House-dove, is rather less.   The Ring-dove is

certainly of greater size, and to it the above remarks will appropriately apply. A Quest which I have just submitted to the scales, in its feathers, as shot, weighs 1¼ lb.; while a Stock-dove put into the same, and under like conditions, barely turns the beam at 12 oz.

In measurement the Quest is 18 inches in length, with a wing spread of 2 feet 5 inches. The length of the Stock-dove is 13 inches, its wings extending to nearly 2 ft. 3 in. By this it appears that the wings of the latter are longer in proportion to its body than those of the former; just what might be expected from its more migratory habits, calling for greater and longer-sustained flights.

Without taking the difference of size into account, the two species, though often confounded by the incurious, are easily distinguished. Though both are of a slate-blue colour, in the Stock-dove the blue is more pronounced; hence one of its common appellations among the country people, of *"blue* pigeon." Nor does it show any white markings, as the Quest, which has these both on the neck and wings. The only variegation on the coat of the Stock-dove—save the lighter and darker shades of slate-blue—is from two or three black blotches (not bands) on its wing-coverts, and the vinous iridescence around its neck, much more brilliant than on the other species, and from which it has its specific name *Œnas* (*oinos,* wine). "Stock" it is supposed to have derived from its habit of breeding in the old stocks of pollarded trees, while the Ring-dove nests higher up among the branches. But there is a more essential difference in their place of nidification; for the Stock-dove does not always make its nest in trees, but rather the opposite. Its hatching-place by preference is certainly closer to the ground, even upon and in it, since it has been known to breed in an abandoned rabbit-hole. But, above all, it prefers the ledge of a cliff, where there is a niche or crevice roomy enough for its purpose. Just such a cliff is there on Peynard Hill, behind my house, where the hard Cornstone overlies a softer stratum of the Old Red; and there the Stock-dove finds the breeding-place of its choice.

This predilection of the *Columba œnas* for rocks has led to its being confused with the real Rock-dove (*C. livia*). Here, in Herefordshire, where the latter is rarer, and of course less known, the Stock-dove is called Rock pigeon, or rather "Rocky,"—when spoken of in the plural number, "Rockies." It is evident that this erroneous nomenclature extended into Hampshire in the days of Gilbert White, and that the "Rockiers" reported to him by the almost octogenarian sportsmen were Stock-doves.

Neither are these last the progenitors of our pigeons, though by a gentle gradation they draw nearer to it. One more link, and we arrive at the real ancestry of the bird for which we provide home and nesting-place. Unquestionably is it descended from the pigeon of the sea-cliffs.

This, the Rock-dove (*Columba livia*), is even smaller than the Stock, and, of course, also the domestic variety; so that "the usual rule of domestication" is not falsified by its being taken as the forefather of the last. The fact that it is so is established by many points of resemblance. The Rock-dove, like the other two wild species, is of a slate-blue colour; but it has the transverse bands upon its wings—wanting in both of these, and always present in blue House pigeons. A characteristic yet more infallible thing shows affinity between the dove of the cliffs and that of the cote—the whitish fleck over the rump, conspicuously seen on both as they spread their wings in flight, but never observed in either Quest or Stock-dove.

But there is no need of this reference to colour for proof of their identity as species. Gilbert White, groping in the darkness of a century and a half ago, found light enough to point it out, when he said, speaking of Sir Roger Mostyn's House doves in Carnarvonshire— "Though tempted by plenty of food and gentle treatment they can never be prevailed on to inhabit their cote for any time, but as soon as they begin to breed betake themselves to the fastnesses of Ormshead, and deposit their young in safety amidst the inaccessible caverns and precipices of that stupendous promontory."

Similar testimony is given by Edwards, the self-taught naturalist of Scotland, who states that House pigeons near the sea-coast in his neighbourhood not only betake themselves to the cliffs, but there interbreed with the Rock-doves, so that it is now impossible to procure one of the latter of pure strain and natural colouring. We have the Rock-dove in Herefordshire. Mr. W. Lloyd, a local naturalist, reports it as breeding on the Stanner rocks, a basaltic upheaval near the border line between the counties of Hereford and Radnor. It has also a nesting-place in the cliffs overhanging the Wye by the celebrated Symond's Yat, and all down through Monmouthshire, to Caldy Island. There, *à fortiori*, they should be found, since these cliffs are nearer to its known habitat on the sea-coast.

The Rock-dove never makes its nest in trees, and is not known ever to *perch* upon them, another point of resemblance to the House pigeon confirmatory of the fact of their having a common origin,

This species also furnishes us with an illustration of Nature adapting her creatures to the mode of life she has designed for them. Its home on the beetling sea-cliff, where it is exposed to the most furious storms, renders it necessary that the bird should be provided with the best means of flight. And just so it is, its wings being longer in proportion to its bulk than even those of the Stock-dove, while its flight is bolder, more arrow-like, and swifter than any of the genus.

The fourth and last species of our native *columbidæ*, the Turtle-dove (*C. turtur*), is also in the list of Wye birds; but only as a summer visitor. This beautiful little creature breeds with us; and its *tur-tur*, from which it has its name, can be heard throughout all the summer's day.

One fact in connection with the Turtle-dove is worth recording. Here, in Herefordshire, its nest is protected from spoliation by a singular sentiment, or rather, super-stition; and nest-robbing boys, who will ruthlessly plunder those of the Ring or Stock-dove, leave that of the Turtle untouched! The reason for thus resisting the temptation, is a belief that any one who robs the nest of a Turtle-dove will—as a consequence, and by way of punishment—soon after have a death in his family!

One day in March my gunman shot four wild pigeons that were feeding on a field of beans recently sown. They do little, if any, damage to the beans at such time; unlike rooks and crows, not "stocking" them up, but only taking those left uncovered, and so lost. It is not about this, however, the present note is written; but to say that, of the four birds killed, one was a Stock-dove (*Columba œnas*), the other three Ring-doves, or, as com-monly called, Quests (*C. palumbus*). They were all in the same flock, which consisted of both species, showing them to associate, at least during the winter, and when after food. This, of course, is nothing new, and I only speak of it to further say that in the Welsh bordering shires the Stock-dove is far from rare, though scarce in comparison with the Ring. In a flock of hundreds of the latter, there may be tens of the former; and he who shot the four abovementioned tells me there seemed about this proportion among those feeding in the bean-field.

Had the aforesaid field been some five miles farther off, on the banks of the Wye, where it *canons* through the carboniferous limestone at Symond's Yat, the Rock-dove (*C. livia*) would, doubtless, have been also in the flock. For there all the three species come together, as it were, on common ground; a singular fact, and of rare occurrence in any other part of the kingdom. Like as not, an odd Rock or two may have been among the feeders in the bean-field, since they sometimes stray a few miles inland from their roosting-place on the river cliffs.

The Rock-dove, so far as I have read, is represented as only inhabiting along the coast-line, nesting in caves and on the ledges of precipices that overhang the sea. I had long suspected that this choice of habitat was not due to any preference for salt water, but merely because the sea cliffs offer the birds better security; and if an equally safe retreat were offered them inland they would take to it. My conjecture has proved correct, and I am now able to affirm that the Rock-dove dwells in the riverine escarpments of the Wye, remote from any sea shore. I have myself noted it as far inland as Hereford-shire; but Mr. James W. Lloyd, of Kingston, an obser-

vant ornithologist, records it as breeding in the Stanner
Rocks of Radnorshire, a trappean formation altogether
away from the sea. Yet Yarrell says: "The Rock-dove,
as its name implies, is a species which in its natural and
wild state inhabits high rocks near the sea-coast, in the
cavities of which it lives the greater part of the year,
only venturing as far inland as may be necessary to visit
the nearest cornfields."

Indeed, reviewing the whole literature of our native
columbidæ, I find it replete with error. To begin with
Bewick, his portrait of the Stock-dove is an excellent
likeness of the Rock, but not at all like the bird it was
designed to represent. Pennant confounded the two
species, saying: "The small sort that is frequent on
most of our cliffs is only a variety of the wild pigeon."
By *wild pigeon*—a very indefinite title—he meant the
Stock-dove, further discoursing of it thus: "The tame
pigeon, and all its beautiful varieties, derive their origin
from one species—the Stock-dove; the English name
implying its being the stock or stem from which the other
domestic kinds sprung." (!) All these assertions are
alike wide away from the truth; for, not only is the "small
sort that is frequent on most of our cliffs" a distinct
species, and no mere variety, but from *it*, and likely it
alone, have descended all our tame breeds. Such, at
least, is the general belief at present existing among
ornithologists. And the name "Stock" has nothing to
do with its being the progenitor of domestic pigeons,
but comes from its nesting in the stocks of old trees.

It is all the stranger that Pennant should have made
these mistakes, seeing that Gilbert White, from whom he
obtained most of his information, evidently knew there
was a specific distinction between the Rock and Stock

doves. Still, the naturalist of Selborne has not spoken
with his usual perspicacity on this point; doubtless, because of the Rock-dove not being a denizen of his neighbourhood—nor yet the Stock breeding there, as he
alleges—he had but slight acquaintance with either.
Montagu also supposed the two species to be the same,
though in the later edition of his "Dictionary," by
Newman, the separation is properly made.

No doubt one of the causes which has led to the two
species having been so often and long confounded is,
that in many parts of England the Stock-dove is called
by the country people Rock, or, rather, "Rocky." It is
so in the western shires, and I think I have discovered
the reason. Instead of nesting exclusively in the old
stocks of trees, as most ornithological writers assert, or
in disused rabbit burrows, as stated by others, it breeds
in cliffs too, on ledges overshadowed by bush or projection of rock. This I can affirm, from having frequently
seen the nest so placed and had the young birds out of
it. Now, as the Ring-dove (Quest), frequenting the same
neighbourhood, never breeds but among the branches of
trees, and the true Rock is usually unknown to them,
this cliff-nesting of the Stock, observed by country
people, would very naturally lead to their giving it the
name "Rocky," to distinguish it from the more common
species, the Quest—just as they have done. Yarrell's
description of the Stock-dove (copied by Mr. Morris,
author of a "History of British Birds," with some slight
alteration of phraseology) is also misleading. He
speaks of its wing feathers, primaries, secondaries, and
tertiaries, being tipped with "leaden grey." There is
not a shade of grey observable on them near the extremi-
ties; instead a dusky brown, at the tips approaching

black. Again, he speaks of certain spots on the wing
coverts, and others on the tertiaries, as of this same
leaden grey. These spots are so near to being black
that no one not colour-blind would think them otherwise,
while those of them described as on the tertiaries are in
reality on the secondaries. As these markings have a
peculiar significance, I will be minute in my account of
them, transcribing from the bird before me—that shot in
the bean-field. There are seven of the secondary coverts
so distinguished; the spots being nearly perfect circles,
and confined to the outer web of the feathers, their edges
quite clear of the shaft. They are of different, indeed
graduated, sizes, and at unequal distances from the tips
of the feathers; else with the wing closed they would
form a " bar," since then only the outer webs are visible.
The other dark markings on the secondaries themselves
—the three inner ones are rather " blabs " than rounded
spots—of indefinite outline—are less conspicuous than
those on the coverts. But between the two sets there is
an evident tendency towards that double oblique bar on
the wings, which makes the Rock-dove so easily identi-
fiable. This is why I speak of these markings as having
a peculiar significance, and in their application to all the
three species of our wild pigeons. It is the more strange,
taking into account their other points of distinction;
their respective sizes, almost in regular gradation from
the great Ring to the little Rock—the Ring without sign
of black mark on the wings, the Stock having them
spotted, almost barred, the Rock with the bars complete!
Besides, the upper tail coverts of the Ring are lead
colour, those of the Stock also, but of a lighter shade,
showing an approach to the white rump so characteristic
of the Rock.

And noting the difference in their habits, we again find a parallelism of gradation. The Ring makes its nest upon the branches of trees, the Stock in the cavities of their trunks, and, as we have seen, also on ledges, while the Rock is exclusively a bird of the cliffs.

Though so near akin and so much alike, nature has certainly adapted each of these birds to a different mode of existence ; but stranger far is the graduated approximation in their habits, combined with that in size, colour, and markings. It is indeed strikingly singular.

HOUSE PIGEONS PERCHING UPON TREES.

On one occasion, while out for a drive, I observed several birds of large bulk perched upon the topmost branches of a tall elm. Their size, shape, and attitudes proclaimed them pigeons, and I, of course, came to the conclusion they were Quests ; but as my carriage came under the tree, which stood by the side of the road, and the birds still kept to their perch, showing no shyness nor sign of alarm, I scanned them more carefully. Wild pigeons, whether Ring-doves or Stocks—Rocks they could not be, roosting on a tree—would not stay such near approach of man—certainly not in this, the winter season.

On scrutiny, they proved to be none of the wild species, but simply House pigeons, that had taken a fancy to curve their claws around a tree branch instead of standing with them flattened out on ridge-tile or cope-stone. There were about a dozen of them, the tree on which they were perched,—seeming perfectly at home

D

upon it,—being close to a house and the cote to which they belonged. Still, not so near but that their thus roosting seemed somewhat strange. I had often seen my own pigeons light upon trees, and, for a time, stay on them; but the trees were in close proximity with their cot, some of them shadowing the gable against which it is fixed. Here it was quite different, the elm being at least fifty yards distant from the walls of the dwelling, and as much more from the outbuildings, where the birds had their home.

This spectacle, so rare, leads to conjectural reflection. Among ornithologists it is the almost universal belief that the domestic pigeon, with all its varieties, is descended from the Rock-dove (*Columba livia*). But this species, so far as I know, never sets foot upon a tree; therefore, why should its tame progeny be doing so? Possibly, and very probably, the answer should be, that the Stock-dove (*C. œnas*) has also had something to do with the progenitorship, this species being in part a tree-percher, but alike a rooster in cliffs; where, as I have lately discovered, it also, and often, makes its nest. Moreover, domesticated pigeons from such paternity would not infringe upon the well-known rule of size-aggrandizement by domestication. With the Ring-dove (*C. palumbus*) it is different; this, the largest of all, having certainly had nothing to do with the procreation of our tame breeds.

———

THE FLOCKS OF WILD PIGEONS.

Another characteristic feature of the Wye-side slopes, at present writing, is the Quest, or Cushat, not as an individual bird, but in grand congregations. The flocks are now at their fullest, and I have never observed them in larger numbers. One I saw this very day could not have counted much less than a thousand. Just now the turnip fields are their special foraging grounds ; and scarce one

NUTHATCH.

but has its little group, if not large assemblage, of these birds moving about among the green tops, which have been prostrated by the late superstratum of snow. As the leaves are rather withered and *délabrées*, the pigeons seem to apply their beaks more to the roots, doing considerable injury to the crop, as the farmer too truly knows. But he has his remedy, since he can recoup himself by shooting them, the Quest not being protected by game statutes. Indeed, the damage they do is far more than made up by the value of their flesh as a food commodity. It is to be remembered that they give some compensation also by the destruction of the seeds and roots of noxious weeds, which would otherwise infest the ground set apart for cultivation.

With regard to the bitterness said to be infused into the flesh of the Quest when fed on turnips, I am still inclined to believe the allegation an error. This very day I have eaten of one in whose crop, when filled, there was nothing but turnip tops, and I am quite sure these had been its food for weeks past, yet I could not perceive the slightest taint of that "rancidity" spoken of by Gilbert White (though not as his own experience), and repeated in almost every ornithological work and cyclopædia written since his day.

The author of " British Birds," in a chapter devoted to the Rock-dove, says : " I have observed in a flock of tame pigeons feeding in a field the hind ones, every few moments, flying over the rest and taking their places in front, to have their turn of the best pickings, and this in constant succession, as if the whole of the flock admitted the right in each other, and claimed it individually for themselves." I think it likely that the Rock-dove acts in a similar manner, but as regards the Ring-dove or Quest, I have never observed it. These certainly do not move so while feeding in the turnip fields, though that is not a true test, since the food thus provided does not call for much moving about. But when they do change place, either walking or on the wing, it is without any regularity of formation or direction. The description, however, if inapplicable to the Quest, is in exact accordance with what I have myself witnessed in the Passenger pigeon of America (*Columba migratoria*). While shooting, or as there called "hunting," these birds in the State of Tennessee, where there are extensive tracts of beech forests, I have seen "gangs" of them so thick on the ground, gobbling up the *mast*, that not a spot of bare earth has been visible between their bodies. Nay, more, they sometimes crowded so close as to alight on one another's backs, as House pigeons may be often seen to do in a farmyard when the food is thrown down to them in a lump. Never stationary, however, these migratory birds of America. With wonderful rapidity those in the advance clear off the fallen *mast*, licking it up, as it were, in an instant, the cohort behind constantly taking wing, and flying over to form the front rank, and so on alternately, till the surface of the ground, or rather its plumed occupants, seem a sea of slate-blue colour, stirred by wavy undulations. I may add that I have discharged a double-barrelled gun, loaded with No. 5 shot, right in the face of such a flock advancing towards me, and at less than forty yards distance, the result, simply to scare them off, without killing a single pigeon. I was never sure about the reason of this failure of the lead to take effect, nor were others to whom the same circumstance had oft-times occurred, the general belief being, that it was due to the wind from the pigeons' wings sending the shot astray. More likely, the thick, close plumage on their gorgets and breasts is the shield which protects them.

The Passenger pigeon is often observed in the northern countries of Europe, and I think it likely breeds in Siberia as well as in America. In the latter, its range extends to the most northern portion of the Continent, and the passage across Behring Straits would be but a few minutes' flight for it. Though having a place in the list of British birds, its claim to this is very slight, resting, I believe, on but a single specimen shot in Fifeshire, Scotland, half a century ago, a waif, in all likelihood, blown over from Russia or Norway.

THE WOODPECKER.

A traveller passing through the shires bordering South Wales, if it be a wooded district, will, every now and then, hear a loud call strangely intoned, resembling, near as may be, the syllables, " glu-glu-glu-gluk," uttered in a sort of laughing giggle. If new to him, it will not fail to excite his curiosity with a vivid desire to know what kind of creature sends it forth. When told it is the call-note of a bird, he will be loath to believe it so; or, if believing, and he has ever heard the cry of the white-headed eagle, he will be half inclined to think it this. But the first rustic met, and questioned about it, will undeceive him, saying: " It's the *heekul*, sir."

He may still fancy the interrogated man means "eagle," with a corrupt pronunciation; and not without further questioning, and some difficulty, will he learn that the loudly-laughing bird is only a woodpecker, little bigger than thrush or starling. Even while he is in the act of inquiring about it, the glu-glu-glu-gluk will again break abruptly on his ear; and if by the side of an orchard, he may see the bird itself flitting from apple-tree to apple-tree, in a pitching, laboured-like flight. Nor does it alight on the branches, but upon the trunk, low down near its base, with head upward, body vertical,

GREEN WOODPECKER.

39

and tail bent inward against the bark, as if to prop it in its place.

Scrutinizing the bird carefully, as he has opportunity now, the traveller will notice that it is of a yellowish green colour all over the back, with a speckle of black and white barring the outer edges of its wings and tail; the crown of its head showing a large, well-defined disc of deepest crimson. If he have visited tropical countries, it will recall to his mind the birds of bright feather he may have seen there. For it is the Green Woodpecker (*Picus viridis*), in splendour of plumage excelling all our native species; perhaps the bee-eater, roller, and king-fisher excepted.

Watching it awhile, after it has pitched against the apple-tree, it will be seen to work upward, not creepingly, but in bold, confident shoots, sometimes direct up the trunk, and sometimes obliquely around it. Now and then it will stop, delve its long, pickaxe-like beak into the bark; and keep delving, quicker than could carpenter or nailer strike with their hammers; its purpose, to lay open the lair of the wood-louse, or insect larvæ, concealed underneath. At intervals, while thus engaged, it gives utterance to its wild, weird cry, which has been likened, and not without reason, to the laugh of a maniac. The note, however, is not always sounded exactly the same: there are times when it is less loud, and softer, and not unfrequently monosyllabic—a single "chook," as if abruptly broken off at seizing or discovering prey. When in full resonance it can be heard distinctly at a mile's distance.

Having ascended the tree to its top, or so far as the bark shows fissures, with the likelihood of creepers underneath, the bird flings off to another, as before, alighting near its base, to repeat every act of the performance.

But the Green Woodpecker does not confine its foraging to trees. Part of its provender it gets out of the ground, ants especially, which I believe to be its favourite food. The length and structure of its tongue would seem to indicate an adaptation for this, the organ being of cylindrical shape, and capable of protrusion fully two inches beyond the tip of the beak. The bird, moreover, has the power of secreting a viscous substance from its throat glands, which, coating the tongue, causes the insects to adhere to it, till they are drawn in between the mandibles and so transferred to the stomach. It is just so with the ant-bears, or ant-eaters, of tropical America and Africa, as also certain other species of birds, formed for feeding on these insects.

While on the ground, the Green Woodpecker progresses in a fashion *sui generis*. Its movements from place to place are made in a series of hops, the head held high, the body erect, as when climbing the trunk of a tree, and the tail slightly spread, touching the earth, not trailing, but as if having a hold on it for the sake of steadiness.

I have had frequent and excellent opportunities of observing this bird's behaviour when after the *formicæ*, and at all seasons, winter and summer. On my lawn, and near the house, these insects abound, so much as to be a troublesome pest, and there the *Picus viridis* often comes in quest of them. In my note-book I find record of several such visits, and during most months of the year; but one paid me in the early part of February, 1879, has attached to it a detailed description of the *modus operandi*. There was a pair of the birds, the Green Woodpecker being of conjugal habits, and as the scene was not twenty

paces from my window, using an opera-glass, I had a
good view of everything. The two were separately engaged, each at an ant-heap of its own. In point of fact,
there was no "hill," the roller having hindered that;
but a slight swelling on the surface told of a colony of
ants underneath in winter quarters—the common yellow
species (*Formica flava*). The bird would plant itself
firmly, with tail hard pressed against the ground, as a
hand to hold by, then commence "stocking," its head
going up and down in rapid repetition, and a ludicrous
resemblance to that of "Punch" in the showman's box.
Thus it would continue, till its beak was buried in the
earth up to the cere, and the head itself out of sight in
the short sward of grass. And while thus it would
pause at intervals, and remain for seconds at a time
without any visible motion, as if *drinking!* What it was
actually doing when thus stationary I can only give a
guess at. My conjecture is, that the tongue was extended
underneath, playing along the hollow passages which the
ants have, and licking up the insects, with their so-called
"eggs"—these last being abundant at that time of year.
No doubt the tongue of the woodpecker is highly sensi-
tive, and its true organ of touch: for its hard, horny
beak cannot be, in this respect differing from the snipe
and other soft-billed *grallatores*.

Notwithstanding the commonness of the *Picus viridis*
in most wooded districts of our country, it is strange
how much of erroneous belief exists about its habits, even
standard ornithologists assigning to it ways it wots not
of, and doings it never did. In a further note I purpose
exposing some of these errors, while further illustrating
the life of this very interesting *avis*.

A BROOD UNDER THE MISTLETOE BOUGH.

In a note published some time ago, I spoke of having discovered the nest of a Green Woodpecker by seeing a litter of chips at the bottom of an apple-tree in my orchard. The cavity in the trunk containing the nest was about seven feet above the ground, and, oddly enough, a fine bunch of mistletoe grew out above, partially over-shadowing its orifice. Standing on tiptoe, and inserting my hand into the hole, there came up out of it a chorus of noises—a jarring and hissing as of goslings, seemingly in anger, and loud enough to be heard full fifty yards off. I say *up* out of it, for the hollow passage, on reaching the heart of the tree, turned downward a foot or so, as I could tell by the direction of the sounds. And that these proceeded from a brood of young birds was equally evi-dent, one of the parents seen near by flitting about among the pear and apple-trees, excited and solicitous.

As the rounded hole was not of sufficient size to admit my hand, I gave up hope of getting a look at the young birds, and turned to note the behaviour of the old one—no doubt the mother. She still kept in proximity to the place, pitching from tree to tree, while every now and then giving utterance to her strange call-note, though in tone more subdued than is usual. And her solicitude seemed less, or at least did not show itself in the fren-zied, distracted way observable among magpies and some other birds, under similar circumstances. Nor did she at any time come very near. All of which I thought strange, knowing the Green Woodpecker to be anything but a shy bird—much less so than either jay or magpie.

Having satisfied myself with watching her movements, I left the place, intending to revisit it on the following day, for further observation, which I did. But on once more thrusting the end of my cane into the cavity, there came forth no noise. All inside was silence, and the birds flown.

Whether the parents had meanwhile carried them off, anticipating my return, with the danger attendant, I am unable to say. For, unluckily, there was another factor in the account, a haymaking boy—we were mowing the orchard grass—with fist smaller than mine, who, in my absence, may possibly have abstracted the chicks. When charged with the theft, however, he stoutly denied it, and all inquiries failed to fix the thief, if such there was. But more likely the young birds had been removed by the mother, as from the time of year (June 29th), and the loud noise they were able to make, they must have been nearly fledged, and so easy of removal.

On having the nest itself drawn out, it proved no nest at all, only some loose "daddocks," as pieces of decayed wood are called by the country people.

A CURIOUS INSTANCE OF SCANSORIAL INSTINCT.

One of the oddest and most interesting habits of the woodpecker tribe is the training their young to climb trees before they are able to fly. At best the woodpecker is a bird of heavy, laboured flight, and often relies more upon its scansorial powers for concealment or escape, than on its wings. Nature has amply provided it with the means for this in the conformation of its feet, claws, and tail, with the guiding instinct as well. But to the latter the parents add instruction, taking the young birds out of the tree cavity when nearly full fledged, and showing them the way to get about. That they do this I have had evidence enough; and a singular case confirmatory of it has just been made known to me by Mr. W. Blake, of Ross, an observant young naturalist, whose word I can well rely on. Out for a walk in the woods, he noticed a hole in one of the trees, some twelve or fifteen feet above the ground, which, from certain *indices*, he suspected to be the nesting-place of a bird. Climbing up to it, he plunged his arm in to the shoulder, to find the cavity turn downwards, and at its bottom felt feathers—a live bird, which, on his clutching it, struggled violently to escape. Drawing it forth, and too much occupied with his own precarious footing, it got out of his grasp, and flew off with a loud, laughing cry, as in mockery—the well-known glu-glu-glu-gluk of the Green Woodpecker. But inside the tree he heard other sounds—the "churming," as he words it, of the young birds; and, re-inserting his hand, he drew them forth one after another—in all five of them nearly fledged. Placed upon the ground at some two or three yards distance from the bottom of the tree, they instantly ran back to it, and commenced climbing up the trunk. They could make no use of their wings, altogether relying on their claws and supporting tail; and with these, doing their best, they soon ascended to the height of six or seven feet—not by a single effort, but several in succession, with pauses and rests between. Undoubtedly, they would have returned to the hole from which they had been taken, but Mr. Blake, having other views, recaptured and carried them home with him.

Two strange facts are exhibited in this occurrence:

first, the parent bird—the mother, of course—having remained within the cavity till caught, a thing so unusual.
There seems no other way of accounting for it than by
the supposition that she was at the time in the act of
feeding her young, and the noise made by them hindered
her hearing and noting the approach of the enemy. That
were intelligible enough ; but the still stranger fact of
the nestlings knowing their way back to the tree where
they had been hatched, would seem one of those instances
of instinct which the philosopher vainly struggles to explain. Unless it were pure instinct, the only explanation
probable is, that they had been out of the hole and down
upon the earth before, while being taught their first steps
in the art of CLIMBING.

In the shires bordering central and South Wales, we
have all four of the accredited British species of Woodpeckers : the Great Black (*Picus martius*), the Green
(*P. viridis*), the Great Spotted (*P. major*), and the Lesser
Spotted (*P. minor*). This might be expected from the
wooded character of many districts in the ancient border-
land of the " Marches."

Of course, the four species are far from being in like
numbers ; the Great Black is so rare that many ornithologists even doubt its existence in any part of England.
It has been observed, however, and in my own grounds
in South Herefordshire, myself the observer. In the
summer of 1880 a pair passed over my head, one flying
behind the other at an interval of a hundred yards or so.
They lit in a tall linden tree near the house, only to stay
in it for a few seconds ; then continued their up-and-
down flight towards some hanging woods beyond, where
I lost sight of and never saw them again. Mr. Chapman
also, curator of the Free Library Museum in Hereford,

records an observation of this species, a single specimen, seen by him on an oak tree in the meadows of Belmont, near the former town. He gives substantial verification of it.

There can be no doubt, therefore, of the Great Black Woodpecker occasionally visiting the Welsh bordering shires, if it be not a permanent resident in them. Against this there is the fact that although these shires abound in woods few of them show timber of large growth; or where it is large the tracts of it are of limited extent. And it is well known that this species specially affects the heart and solitude of the thick forest, rarely coming out into the open; while with the other three it is different. In this retiring habit of the Great Black Woodpecker I note a resemblance between it and the two American species most nearly akin to it in size as in colour, both being black. I mean the Ivory-billed (*Picus principalis*) and the Black or "log-cock" (*P. pileatus*). These always keep to the interior of the grand primæval woods; their loud tapping — from which they have derived the fanciful name of "carpenter birds," in Spanish America, *carpinteros*—and their still louder call-note, oft startling the traveller, as he rides silently along some lone, shadowy aisle of the forest. And on the other side of the Atlantic, just as on this, the smaller and spotted, or mottled species—of which there are several—more affect open woods, some of them frequenting orchards, and nesting near the homestead.

Taking our English woodpeckers, not in the order of size, but scarcity, one or other of the two so-called "Spotted" species claims attention next, though it is difficult to determine which. Both may be pronounced rare birds, and are so compared with the Green Wood-

pecker; in many districts one or the other altogether
wanting, and some where neither is known to exist.
From my own observation of them in Herefordshire, a
balance might be struck as to their abundance, some
"lays" of country seeming to have more of the Great
Spotted, others of the Lesser, just as in one place there
will be only tree pipits, while in another, near by, those
of the meadow species alone are seen.   Perhaps as good
a guide as any to the comparative numbers of the Great
and Lesser Spotted Woodpeckers, taking them all over
the country, is to be found in the price lists of the
taxidermists.   One I have before me gives the following
quotations :—Skin of the Great Spotted, 3*s.* ; egg, 9*d.*
Skin of the Lesser Spotted, 3*s.* 6*d.* ; egg, 2*s.*

By this it would appear that the Lesser Spotted is the
rarer bird, and its nest more difficult to find.   Still, that
may arise from its more diminutive bulk, making it less
conspicuous and so less liable to be shot.   Certainly in
my neighbourhood, and about my own grounds, it is the
more plentiful of the two, as also throughout the adjacent
Forest of Dean, where both species are met with in considerable numbers, though still far from common.

The name "Spotted" is for either much of a misnomer.
There is scarcely a spot on them, but instead several well-
defined bars of black and white, so that "barred" would
be a more appropriate appellation.

From my observation of the two species, their habits
appear to be much alike, while differing in many respects
from those of the *Picus viridis.*   They keep more within
the woods and to timber trees than it—especially the
Great Spotted ; while the Green is a forager in orchards,
and alights on pasture fields where ants, left long undisturbed, have thrown up their hundreds of hillocks.   I

have never seen either of the others "stocking" at them, though the Green so engaged is an every-day sight.

About the "tapping" of these birds, so much talked of as to be the burden of many a song—notably that sentimental lay of poor Henry Kirke White, who was wrong in making it a beech tree—I believe this noise is oftener made by the Nuthatch than any of the woodpeckers. The Lesser Spotted certainly gives out such a sound while searching for its food, but it more resembles a "whirr" than tapping, as a piece of dry stick drawn rapidly across a coarse-toothed comb. The other two also "tap," while splitting the bark to lay open the lair of the woodlouse; but the sound made by them is not perceptible at any very great distance.

The voices of the two Spotted species, so far as I have heard them, are quite different from that of the *Picus viridis*, differing also from one another. That of the Great Spotted is a monosyllabic note, a "chuck" very much like that the starling sometimes utters, repeated at intervals of nearly a minute each; while the call-note of the Lesser comes nearer to that of the Green Woodpecker, only of fainter, feebler tone.

Of all the four British species, the Green Woodpecker is the one of commonest occurrence, and so best known. Still, its habits are less understood than might be supposed, some of them even being incorrectly described by ornithologists of greatest note. In a future chapter, to be devoted to this interesting bird, I purpose rectifying such of these errors as I have found the facts to contradict.

SOME FALLACIES RELATING TO THE GREEN WOODPECKER.

One among the many curious habits ascribed to this bird, in common with our other species of Woodpecker, is, that the jarring noise made by it on the bark of trees is a signal of communication between the sexes! Singular enough, were it true, which, in my opinion, it is not— instead, only a tale worthy of the credulous Pliny, or the romancing Buffon. Yet Montagu not only believed it but of himself has absolutely affirmed its truth, as follows:—"The jarring noise so frequently heard in woods in the spring is occasioned by *one or other of this genus,* which, from frequent observations, we have no doubt is used as a call by both sexes to each other. It is curious to observe them try every part of a dead limb till they have discovered the most sonorous, and then the strokes are reiterated with such velocity that the head is scarcely perceived to move, the sound of which may be distinctly heard half a mile."

Now, if Col. Montagu, while listening to this peculiar noise, saw the bird which made it, why was he unable to tell its exact species? The words I have italicized clearly show his uncertainty in this respect; for to such an accomplished ornithologist a glance should have been enough to distinguish the Green Woodpecker from either of the spotted kinds. If ignorant even of the bird's identity, it seems a stretch of imagination on his part to endow it with a habit, or instinct, so extraordinary— indeed, outside nature. Surely *she* provides for all her creatures the means of communicating with one another by their own organs, without the necessity of resorting to extraneous instrumental aid. I cannot think of one that does this; though there may be, and is, if the ticking of the "death-watch," as entomologists assert, be a call-note to its mate. But why should a Woodpecker, with enough volume of voice to make itself heard to the distance of a mile—why should it, of all others, employ a bit of loose tree-bark as a sound-board in the utterance of its amorous speeches?

The truth seems to be, that the resonant bark, being hollow underneath, affords shelter to the woodlouse, with other prey of the *Picidæ*, these knowing that the noise will start the insects out, and so spare them the labour of hacking and splitting.

Col. Montagu further contradicts the statement of Dr. Plot, that the tapping noise, usually attributed to Woodpeckers, is produced by the Nuthatch. Yet the doctor was in the main right, the colonel altogether wrong.

Yarrell, who, I believe, still stands at the head of British ornithologists, has also made mistakes about the habits of the Green Woodpecker. He says, "It is one of the earliest birds that retire to rest in the afternoon;" whereas *it is one of the very latest.* Scores, hundreds of times, have I heard its loud "cackle," and seen the bird itself flitting about my grounds till the last gloaming of twilight.

Again Yarrell states, this time on hearsay authority, that Green Woodpeckers "when excavating a hole in a tree, for the purpose of incubation, will carry away the chips to a distance, in order that they may not lead to a discovery of their retreat." Wise birds, were it so! Which it is not; instead, the very reverse, as I have ample evidence, the " chips " often betraying the *locale* of their retreat, or more properly speaking, their nest. An instance in point once occurred to myself, when I discovered the nest of a Green Woodpecker by seeing a quantity of whitish-coloured fragments scattered about at the bottom of an apple tree in my orchard—in all, over a quart of them. Divining their origin, I looked up the trunk, to see, at about seven feet from the ground, a cavity with circular orifice, unmistakably the nesting-place of *Picus viridis*—which on examination proved to be the case.

Another fanciful belief about the Green Woodpecker,—so common as to have earned for it one of its trivial appellations,—is also mentioned by Yarrell, who says: "They are said to be vociferous when rain is impending, hence their name of rainbird." He thinks this probable, and offers scientific explanation of it, in the feathers of birds being readily affected by electricity, and so forewarning them of changes in the weather. I have heard the Green Woodpecker sounding its note throughout the morning hours when there was neither cloud nor other sign of rain in the sky; yet in the afternoon came a downpour. Therefore I, too, might have believed there was a connection between the bird's call and the condition of the atmosphere, but for hundreds of other instances contradicting this idea. Many a time and oft have I listened to it laughing its loudest, and for days in succession, during which not a drop of rain fell —at times, too, when this was much wanted. While the rain is actually falling, then the bird is usually vociferous enough; but that is not prediction; more likely delight at thinking the deluge of water may drive out ants and other insects from their places of concealment.

True, there is nothing very improbable in this bird, as many others—beasts as well—being in some mysterious way forewarned of approaching changes in the weather. I only know that the warnings it is itself said to give, by a call of especial loudness, are not to be relied on; and when heard preceding a rainfall it seems simply coincidence.

I think, then, the above beliefs have been shown to be more or less fallacies, notwithstanding their having been religiously copied by Mr. Morris in his "History of British Birds," and by a host of other writers—in short, they have run the rounds of most ornithological works, including encyclopædias, and are still running them.

Yet another of these fancies, though less worthy of note, may be alluded to: that of the Woodpecker "tapping at the hollow beech tree," a conceit, no doubt, originating in the brain of Henry Kirke White, and perpetuated by his gentle lay. It would be a rare sight to see a Woodpecker on a beech tree, whether hollow or sound, for the simple reason that the bark of these trees is seldom otherwise than sound, affording no lodgment to insects, besides being too smooth even for the claws of the *Scansores*. The apple tree, knotty, corrugated, and swarming with insect larvæ, is the favourite habitat of the Green Woodpecker; and, no doubt, the abundance of this species in the "cider shires,"—greater I believe than elsewhere,—is owing to the orchards.

Elsewhere I said that the "tapping" oft heard in woods is more the work of the Nuthatch than of any of the species of *Picus;* and I now find, on referring to "White's Natural History of Selborne," that he pointed out this fact more than a century ago. Indeed, the Green Woodpecker, which, as the largest of the three common species, and, armed with the most powerful beak, might be expected to make the most noise in this way, scarce makes such noise at all. Neither does the Greater Spotted; while the sound proceeding from the Lesser Spotted is unlike that produced by the Nuthatch, and nearer to the "skirr" of a rattlesnake.

The ordinary note of the Nuthatch bears resemblance to the twittering of swallows, but fuller in tone and louder. What may be called its song, however, is a sort of piping strain, rather sweet, but peculiar for the voice of a bird, and bearing resemblance to the sounds produced by the little water-whistles known as "nightingales."

THE NUTHATCH.

In one of his letters, bearing date April 18, the naturalist of Selborne says:—" Now is the only time to ascertain the short-winged summer birds; for when the leaf is out there is no making any remarks on such a restless tribe; and when once the young begin to appear it is all confusion—there is no distinction of genus, species, or sex." Taken literally, the above might lead to erroneous inferences; but the meaning is, of course, clear, Mr. White intending to point out the great difficulty encountered in the observation of birds, and their habits, during the time of year when the trees are in full leaf. He seems to refer only to the birds which are our summer visitants; but his remarks will equally apply to many of the species permanently resident; such as during the winter are shy and keep far afield, so giving less opportunity for observing them. Among these may be mentioned the Nuthatch (*Sitta Europæa*), which in early spring more frequently enters the orchard to forage after the flower buds of plum, cherry, and other stone-fruit trees. Less shy at this sea-

son, it permits nearer approach, and so can be better seen
and its habits observed. I myself look upon the Nuthatch
as one of the most interesting of our native birds ; for it
is truly a native, not only nesting with us, but remaining
throughout the year. Part of the interest attaching to it
is the peculiar position it holds in our ornithological list, it

NUTHATCH AND JAY.

being the only species of its genus which either inhabits
or visits the British Isles, while the genus itself is marked
by many peculiarities. Its rarity may be also said to con-
tribute to its attractiveness, as with almost everything
else. For although in wooded districts it cannot be
called uncommon, it is nowhere very numerous, and from
many neighbourhoods altogether absent. Independent

of all the above, it is a remarkably handsome bird when
in perfect plumage, which, though neither so brilliant as
that of the Jay or Green Woodpecker, is nevertheless
aught but sombre. A specimen (stuffed) I have before
me, which was shot in my orchard last winter, shows the
back and upper parts of the body of a light slate blue,
the breast, belly, and under parts a bright though delicate
buff; while the under tail inverts, and feathers around the
vent, are a rich ferruginous red.

Still more interesting are the habits of the Nuthatch,
so widely differing from those of our other *Aves*. Its
being able to run up the trunk of a tree shows relationship with the genera *Yunx*, *Picus*, and *Certhia*. But it
is a better climber than any of these, the last, perhaps,
excepted,—since it can run down as well as up, and this
notwithstanding that it lacks the stiff supporting tail
feathers, which the creepers and woodpeckers have. Like
the wryneck, it has no prehensile power in its tail.

The name *Nuthatch*, synonym of nut-hack or nut-hacker,
is perfectly appropriate. Some days ago one was seen
in my orchard on a large limb of a *Bon-Chrétien* pear
tree, busily jobbing away at something on the branch
before it. The strokes were delivered in rapid repetition,
and so loudly as to be audible at more than a hundred
yards distance. Thus occupied, it permitted near approach; so near, the observer had no difficulty in noting
every movement. He could see that the beak was driven
down, with the head at times held a little sideways, while
at each dig there was a muscular straining of the legs, as
if to give better force to the blow. After a time it flew
off, bearing between its mandibles what looked like a
piece of bark; but it was more probably the kernel of a
nut, or some other edible substance. On the observer

ascending to the branch where it had been at work, he
found a fissure in the bark; which, no doubt, the bird
had been taking advantage of to hold the object it was
hammering at.

On one occasion a bird was brought to me for identification by a ranger of a neighbouring wood. He had
shot it, not within the wood, but beside it, in the garden
of his lodge, where it was feeding upon the young flower
buds of a cherry tree, not yet blown. I saw it was the
Nuthatch (*Sitta Europæa*), whose favourite food is the
hazel nut, from the breaking open of which with its
powerful pickaxe beak it derives its vernacular name—
presumably an altered form of "nuthatchet," or "nut-
hack." Failing the hazel nuts, it will eat acorns, beech-
mast, berries, and the kernels of stone-fruit, as also
beetles and other insects, though I think it prefers a
vegetable diet when such can be obtained. The fact of
its being taken in the act of despoiling the cherry tree is
somewhat confirmatory of this; for although strictly a
wood bird, and commonly confining itself to the timber
trees, there are certain periods of the year, as now in
early spring, when it pays a visit to the adjacent orchards
to make forage among the buds and blossoms.

The Nuthatch is fairly entitled to a place in the list of
interesting British birds, and for several reasons. In
addition to its very pretty plumage, it is the only species
of its genus we have; while its habits are singular and
sui generis. Besides, it is of somewhat rare occurrence,
for although inhabiting many wooded districts of our
island, it is far from being common, and still farther from
being commonly seen. Even in the neighbourhoods it frequents but few people are acquainted with its *personal* appearance. As a proof, the man who brought me the speci-

men knew not what bird it was, though he has been rang-
ing the woods around for upwards of thirty years. Yet in
these very woods Nuthatches are perhaps as numerous as
in any other part of England. This good man, however,
is not given to ornithological observation. His business
is with timber, lop and top, the split laths for palings,
the hurdle-bars, hop-poles, and pea-sticks. All these he
thoroughly understands, from the cutting down to the
carting off after being sold, and the price a purchaser
ought to pay for them. But birds he knows nothing
about, neither does he profess it. Alike deficient is he
in a knowledge of the four-footed *feræ naturæ*, and equally
candid in disclaiming it. I verily believe that, while
going his rounds, if an eagle flew over his head, or a wild
cat scampered across his track, he would think no more
of the first than if it were only a sparrow-hawk, and as
little of the last as though but a rat or weasel. As he is
in every sense an honest, respectable man, I can forgive
him for this absence of interest in things which so much
interest me; though as a study to the naturalist—for man
is no exception to the subjects with which natural history
has to deal—his proclivities are as much a puzzle to me
as the mode in which the cuckoo deposits her egg in a
nest too small to admit the possibility of her there laying
it, or the manner of procreation ascribed to the vivip-
arous blenny.

But I must leave the unobservant wood-ranger, and
return to the bird of whose species he was ignorant,
though it must have flitted before his eyes some hundreds,
if not thousands of times. The Nuthatch is deserving of
notice from the naturalist, much more than it appears
ever to have had. I have pronounced its plumage pretty,
and, without entering into minute details of its colour or

markings, it may be described in general terms as half leaden-blue, half buff. The blue is above, comprising the crown of the head, the nape of the neck, and back; the buff below, taking in the throat, breast and belly, the general tint of the under parts showing an admixture of chestnut and orange. A black list runs from the base of the bill over the eyes and on to the shoulder. This mark has a peculiar meaning, as I shall presently show. The long, strong, conical, and sharp-pointed beak is dark blue above, the convex ridge of the lower mandible being of a whitish horn colour. Morris, in his book of "British Birds," describes the legs, toes, and claws as *brown.* In the specimen before me, neither the legs nor toes are of this colour; instead yellowish-red, with the same slight admixture of orange observable on the plumage along its sides. The bird is six inches in length; but the tail being short in proportion to the body allows for a greater bulk than might be deduced from the measurement. It is, in fact, about the size of a greenfinch, though of quite a different shape, in form more resembling the woodpeckers. To these it is also very similar in habits; and although classed with the *Certhiadæ,* or Creepers, its affinity to the *Picidæ* seems quite as close, or closer. Its resemblance to the woodpeckers is noticeable in many of its ways. Like them it is a true tree-borer, not only delving into the bark after insects, but drilling a hole for its nest. The noise it makes while engaged in this operation can be heard at a considerable distance, and is often mistaken for the "tapping" of the woodpecker. A somewhat similar hammering is made by it in breaking open nutshells to extract the kernels; from all of which it has obtained the additional titles of "nut jobber" and "wood-cracker."

Another point of similarity to the woodpeckers, not in habits, but in plumage, is the streak or list already alluded to as running longitudinally from the base of the beak over the cheeks and on towards the shoulders. This moustache-like marking is a peculiar characteristic of all the woodpecker family, and seems to have some mysterious connection with their mode of life. It is, at least, strange that the Nuthatch, of such similar habits, should also be thus similarly provided, the thing itself pointing to an alliance between the two *genera*. The *Sitta Europæa* is a true tree *climber*, or rather *creeper*, since its mode of progression is that distinctive of the *Certhiadæ*. While moving upon the trunk or along the larger branches, it does not hop as the woodpeckers, but walks foot over foot, in a quick, stealthy gait, its body flat against the bark. Nor does it assist itself with the tail, of which the woodpeckers make much use as a prop and support, often even when they are upon the ground. Moreover, these seem only able to go *up* the tree, or around its trunk, while the Nuthatch can "swarm" with equal facility either upward or downward. What gives it this superior capacity will be apparent by an examination of its foot; the hinder toe, or heel, being larger than any of the three anterior ones, while all are furnished with large sickle-shaped claws, sharp-pointed, and strongly prehensile. If the top of a finger be inserted between them and rapidly drawn forth again, they can be felt adhering to it as though they were barbed. From this it may well be supposed that the slightest inequality in the bark will be caught and clutched, without danger of the bird slipping off, whether head up or down.

As already hinted at, the Nuthatch nests in a tree cavity, in this respect also as the woodpeckers. And

like these, it delves its own hole, though sometimes
taking possession of one already hollowed out If the
aperture of this be larger than is necessary for the admis-
sion of its body, the bird has been known to make it
narrower by laying a plaster of mud or clay around the
orifice.   This trouble is taken, suggests Yarrell, as a pre-
caution against attacks by the tits, a small embrasure
being easier of defence than a large one.   The reason is
rather unsatisfactory.   A blow from the powerful beak
of a Nuthatch would send tomtit, even the great *Parus
major*, to perdition.   More likely the "chinking" is
done to hinder the entrance of hawk or owl—possibly the
pole-cat.   When the Nuthatch excavates for itself, the
hole is a cylindrical tunnel, first running horizontally,
then at the end dropping downward to the site of the
nest,—a loose deposit of leaves, bits of bark, and moss,—
where it lays six or seven eggs of a dull white colour,
spotted, or blotched, with brown.   In the pairing season
its note, "kweet-kweet" may be heard, though at other
times it is rather a silent bird.   Its presence is more
often betrayed by the noise it makes while hammering
at the hazel-nuts.   Its mode of extracting their kernels is
perhaps the most curious thing relating to it.   In order
to keep the nut steady to receive the stroke of its beak,
it first presses it into a crack of a decayed tree, or a
crevice in the bark,—sometimes between the posts of a
paling,—just as a blacksmith fixes in his vice the iron he
intends operating upon.   And while pecking at the shell
the bird is so well sustained by its claws as to have the
whole body at command, which moves up and down with
the blows, its weight giving strength to the stroke.
Take it all in all, the Nuthatch is one of the most interesting of our indigenous birds, for it is a true native,

nesting with us, and continuing its sojourn throughout
the whole year.

Mr. Brammer, one of the wood-wards employed in the
adjacent Forest of Dean—a Government property—tells
me of a bird which makes its nest in a very original and
singular situation.   When a portion of the Forest timber
is cut down, for the slabs and props used in the coal-mines,
it is first stacked or corded, the "cords" being separated
by upright stakes driven into the ground between.
When the wood is hauled away, these stakes are often left
standing, and remain so for many years.   After a time,
the weather having free play upon them, they become
partially decayed ; and then a small bird,—a *tit*, as my
informant supposes it to be,—hollows out a cavity in one
or other of them, near their top or head, in which it makes
a nest and brings forth its young.   A small round hole,
he describes it, running several inches into the stake,
horizontally at first, then lowering to the nest.   Mr.
Brammer, although a truthful and intelligent man, is, like
my nearer neighbour, the ranger, not much of a naturalist ; and I take it that his "tit" is neither more nor less
than a Nuthatch.   At all events the bird certainly does
not belong to the family of the *Paridæ.*   For, though the
latter often make their nests in holes of trees, they do
not themselves make the holes, and cannot.   I intend
paying a visit to these timber *troglodytes,* and scraping
acquaintance with them.

THE SCARCITY OF SONG THRUSHES.

I have never known Song Thrushes so scarce as they are at present, and have been during all the past year, 1880. I speak of my own neighbourhood, South Herefordshire, though I have reason to believe it is the same all over the country. Three summers ago, in my grounds, I could hear two or three of these birds of song,—unmatched, save by the nightingale,—singing at the same time, and within a stone's throw of one another; and singing all day long, from early morn till dewy eve, so constantly and continuously I often wondered at vocal powers that seemed never to fail or flag. But now all is changed, and so changed! The mellifluous notes of the mavis are rarely heard; and when heard it is in solitary strain—but one bird singing within earshot, and that only on occasional days. Nor is this the worst or strangest part of it—still another change seeming to have come over the thrush, making it *parsimonious* of its song. Instead of the prolonged strain of former days, this year, whenever and wherever I have heard it sing, there was but the going over of its gamut two or three times, and all silence for hours after!

This fact, for it is a fact so far as my observation extends —and I have several times observed and been surprised at it—courts inquiry as to its cause. Can it be because the thrushes are so few in number, each pair with a wide field to themselves, that the cock bird, having no rival near, and therefore no motive to make display of his pre-eminence in song, is for this reason so sparing of it? The conjecture that such is the cause may seem ludicrous —yet I can think of no other. And why may it not be thus? It is well known that caged birds sing better in

company ; piping out their notes in jealous rivalry, as
human vocalists on the stage of the opera-house or concert-
room.   And why not wild ones the same ?

PROOF POSITIVE OF THRUSHES BEING SCARCE.

If, beyond the facts above set forth, I had needed other
evidence to assure me of the Song Thrush being now in
diminished numbers, I have got it in a way convincing,
as curious.   Some days ago, chancing to be within earshot of two boys, one of them the most noted nest-robber
of my neighbourhood, I overheard a snatch of dialogue
to the following effect :—

"Wonderful few o' the singin' Thrushes be about this
year, Jim."

"What make 'ee think that, Dick ? "

"'Cause I hain't foun' a nest o' em yet, an' there warn't
a many last year, eyther."

"Theer be plenty o' the mistletoes ; more'n I've ever
seed.   I hear 'em screechin' all about farmer's big
orchard."

"Oh ! bother the mistletoes.   They bean't much good ;
neyther their eggs nor themselves.   But the singers !
If I only had a nest o' young 'uns now, I could get five
shillin' for 't."

Dick was the famous bird-nester ; and at this point,
discovering myself, I interrupted the dialogue.   I called
him up, for a spell of cross-questioning.   Submitting him
to this, I found he was fixed in his idea that the Song

Thrush was less numerous in the neighbourhood than it used to be, even within his brief period of nest-plundering existence, though he was unable to assign the reason for it. This set before him, as proceeding from the severe winters of 1879-80 and 1880-81, he caught the idea up, instantly exclaiming, "That's it sure, sir. I knows the singin' Thrushes be wonderful *nesh.*" By the old saw, there are "sermons in stones, and books in running brooks," and just such teaching got I from this ragged boy, though the lesson was but confirmatory of my own observations, already made.

THE MISSEL THRUSH ABUNDANT.

The conversation which is above reported gave hint of another fact, worthy of a word or two, and one I had also been speculating upon. This, that the Missel Thrush, by the boys termed "Mistletoe," is in as great numbers as ever, if not greater. This would accord with the ornithological character of the bird, in connection with the peculiar circumstances which have marked the two winters spoken of—both severe beyond the common. The Missel Thrush is a much stronger and hardier bird than the mavis, and will even outlive winters that kill the fieldfare and redwing, two congeneric species, which one might suppose, by their breeding and spending the summer in more northern climes, would be better able to endure cold in its extreme degree. Still, I believe it is not the cold which tests the strength and endurance of these birds,

F

but hunger ; and very likely the Missel Thrush, to the manner born, and able to subsist on mistletoe berries,

MISSEL THRUSH.

with those of the ivy and others the snow cannot all cover up, is thus preserved in undiminished numbers.

CHAFFINCH, OR BACHELOR BIRD.

Fringilla Cœlebs—Bachelor Finch—the name which Linnæus bestowed upon the Chaffinch, is a misnomer—at least, in Siluria. The Swedish naturalist has said that "before winter all the hen Chaffinches migrate through Holland into Italy." The remark, of course, refers to Sweden ; but commenting upon it, the famed naturalist of

Selborne says :—" I see every winter vast flocks of hen Chaffinches, but none of cocks."

Now, I have been observing the Chaffinch, one of our most familiar birds, for several years throughout all the winter and summer, and have never known the sexes so to separate. In all cases where there were flocks, the cocks and hens seemed to be in about equal numbers, or at least no difference worth noting; and Mr. Knapp, the author of "The Journal of a Naturalist," bears similar

CHAFFINCH.

testimony of them. He says, " With us the sexes do not separate at any period of the year, the flocks frequenting our barn doors and homesteads in winter being composed of both." Mr. Knapp's observations were made in Gloucestershire on the left bank of the Severn; mine chiefly in the valley of the Wye. So, if those of Linnæus and Gilbert White be correct, then the habits of the birds in these western shires must differ from what they are elsewhere, even in our own islands—a somewhat singular circumstance.

It is not often that the amiable naturalist of Selborne came to wrong conclusions, or put forth fanciful theories, so commonly indulged in by writers on natural history. He was too acute an observer for the former, and too conscientious a one for the latter. Yet in regard to these same birds, another of his ideas seems paradoxical—that relating to their migration. Noticing the large flocks of them that appear in winter, he says, " It would seem very improbable that any one district should produce such numbers. . . . Therefore we may conclude that the *Fringillæ cœlebes* for some good purposes, have a peculiar migration of their own." Now, when we consider that the Chaffinch usually produces two broods in a year, each of four or five birds, and that around every house, and in almost every hedge-row, there is a nest, it is mere matter of wonder the winter flocks are not larger than they are. Certainly migration is not needed to account for their numbers. And, possibly, there is a like easy explanation of the Hampshire ones being *"almost* all hens," as White puts it. For he does not affirm that they were *all* hens. May not the predominance of this sex have been only apparent from the young cocks of the year not yet having attained their perfect plumage—the red breast and brighter hues generally ? Might not these have been mistaken for hens, and so made the latter appear the more numerous? Supposing eight or ten young birds to be successfully brought up by a single pair in the breeding season—and admitting the above theory to be correct—these, with the mother hens, would give in the winter flocks nine or eleven grey breasts against one of the brick colour. And, like enough, this is the explanation of the puzzle.

THE BACHELOR BIRDS.

In the valley of the Wye no species of our smaller birds is represented by so many individuals as the Chaffinch (*Fringilla cœlebs*). In a miscellaneous flock, congregated in either field or farm-yard, composed of buntings, sparrows, linnets, greenfinches, and Chaffinches, these last will usually outnumber all the other kinds; in rare instances only, and in certain spots, the sparrows mustering in equal strength. But in Herefordshire, throughout all the year, winter or summer, the Chaffinch is the bird most familiar to the eye, ever present to the sight, whether the spectator be journeying along the road, sauntering through the fields, or looking forth from the door of his dwelling. Its somewhat monotonous, yet still cheerful, "twink-twink," salutes the ear with like frequency; though this is not audible at all seasons, since the Chaffinch, in common with most other birds, is mute during the chilly days of midwinter. This winter it has been so for a much longer period than is its wont. Its song, not unfrequently heard about the middle of January, did not strike my ear till February 6th, after the thaw had declared itself, and the thermometer run up to 45°. Nor till that time did it sound its ordinary call-note. Now, both call and carol enliven the copse, and ring around the walls of the dwelling. The song will again cease about midsummer, but not the twink-twink; that will continue on till the cold of the autumn once more admonishes it to silence.

Linnæus bestowed upon this bird the specific name *Cœlebs* (Bachelor), because, as he says, the sexes at the approach of winter become separated; adding, "All the hen Chaffinches migrate through Holland into Italy." Of course he speaks of a migration from his own country, Sweden. Gilbert White, referring to the same bird and its habits, as observed by him in Hampshire, after a fashion confirms the statement of the Swedish naturalist. He says, "Vast flocks of hen Chaffinches appear with us in winter, without any cocks among them." Such partition of the sexes does not take place here in Herefordshire; at least, it has not come under my observation. Nor does it in the north of Ireland, where, in my earlier days, I was well acquainted with the habits of the Chaffinch, there erroneously called Bullfinch, or still more erroneously, "Bullflinch." According to Mr. Knapp, author of "The Journal of a Naturalist," neither is there such a separation in the adjoining county of Gloucester. So far as I have seen, in all the flocks frequenting this neighbourhood for several winters back, the two sexes have been in about equal numbers; and where only three or four birds are seen together, one or two of them will be red-breasted. Morris, in his book, "British Birds," while chronicling the circumstance of the sexes so keeping apart—which he believes to be a fact—says : "I am inclined to think that this is most frequent in severe winters." My experience of this winter on the Wye falsifies this conjectural assertion. It has been one of the severest; yet throughout its severity the flocks of Chaffinches have been composed of males and females, as many of the one sex as the other.

There are people who speak of the Chaffinch as an uninteresting bird, an assertion showing little of either sense or taste, and an opinion with nothing to support it, save it be the plenteousness of the creature so harshly judged. Were Chaffinches scarce with us as Java sparrows, no doubt they would be more appreciated, and, like the last, oftener confined in cages. Luckily for them they are not of such rare occurrence. The male Chaffinch, the "Bachelor," is in reality a beautiful bird, his plumage of the very brightest and gayest in our indigenous aviary. And the female, too, though of hues more sombre, when closely examined, shows shades and markings becomingly pretty. To speak of any bird as uninteresting is to give utterance to the language of a Goth; above all, as regards the *Fringilla cœlebs*, which in the drear winter day cheers us by its ever-presence, coming close up to window-sill and doorstep ! As well might one say wicked things of another red-breasted bird—the Robin; and none will dare do that.

THE BACHELOR BIRD A FRIEND TO FRUIT GROWERS.

Many an anathema is hurled at the head of the Chaffinch, alike by farmers and gardeners; and too often a shot from the ten-shilling licensed gun. Nor can it be denied that *Fringilla cœlebs* does damage to the young sprouting wheat, and the seedlings of the kitchen garden. But let justice be done to the bird, and account taken of the compensation given by it in the destruction of noxious *larvæ*, feeders both upon fruit-tree leaves and those of garden vegetables. Just now, it so happens that apple trees are infested by a "blight," of quite unusual severity, causing great anxiety to fruit growers, these hideous grubs doing great injury to the trees. Often the hopes of a whole orchard, about declaring themselves in full bright promise of blossom, are crushed—as it were, literally nipped in the bud—by them. And this very year there is every appearance we shall have a shortening in the fruit crop, if not actual failure, from the same cause. It will be less, however, in an orchard where Chaffinches abound, as these birds, now with young in the nest, are industriously collecting caterpillars from the apple and other trees, to supply the stomachs of their broods, like Oliver, ever calling for more.

I can certify to this beneficial fact, from having been an eye-witness to it day after day. Therefore I would be-seech the destroyers of small birds to show mercy to the Chaffinch—if only for the sake of their pears, apples, currants, and gooseberries.

A CHAFFINCH PARTIAL TO NEWSPAPERS.

Though in building their nests each species of bird employs certain materials by preference, yet, as is well-known, where these are wanting, birds will use such others as come nearest the thing of their choice. Few make a neater nest than the Chaffinch, and it is rare to find one greatly differing from another. Yet I have a Chaf-finch's nest now before me, which displays eccentricity of a somewhat comical kind. Instead of the lichen usually enamelling the outside, this is mottled all over with bits of newspaper of different sizes, neatly worked into the wall of grass work and other materials. Examining a number of these scraps, I find them chiefly taken from the advertising columns; though no doubt the bird intended them for concealing its habitation, rather than making it known to the public.

This nest was found in the shrubbery of one of the *town* gardens in Ross, where lichen may have been scarce, while scraps of old newspapers lying about in plenty served the bird as a substitute. Withal it is rather an odd case of accommodation to circumstances.

CHAFFINCH AND CHIFF-CHAFF.

Throughout the month of May and up to June these two birds are heard almost continuously from earliest daybreak to latest twilight. The ordinary note of the Chiff-chaff, which resembles the sound made by the file in sharpening a saw, is anything but agreeable, many people pronouncing it the reverse; while the strain of the Chaffinch, though cheerful enough, becomes tiresome through constant repetition. One day I took out my watch to time one which was singing in a tree close by; and, after carefully counting, I found that it repeated its song 7½ times to the minute, or 450 in an hour. And for many hours of the day this was kept up, with only now and then short intervals of silence. We could forgive the "Bachelor bird" for the plenteous outpouring of his monotonous note, as it cheers us at a season when most other songsters are chary of theirs, or altogether silent. But it is withal somewhat vexatious just now, as it hinders the hearing and distinguishing the songs of rarer species, who make but a short stay with us.

EARLY APPEARANCE OF THE CHIFF-CHAFF.

Having read various accounts of this summer visitant being seen in the month of February, I was disposed to doubt the correctness of the observation, and so said in a former note. I now withdraw my doubts, and make apology to the discredited observers, having myself shortly after seen the Chiff-chaff, and held it in my hand. Still, I cannot think that these birds have come on the regular return migration from the South, but have been staying with us all the winter. Why should they not any more than siskins, gold-crests, and other species seemingly tender as they? It is quite possible, even probable, that many Chiff-chaffs remain in England throughout the winter—when this is mild—and are not noticed. For then not uttering their odd repetitive note, they might easily escape observation, or, if observed, be mistaken for other species. The theory held by some people, that this bird hybernates—by which I suppose they mean that they lie up somewhere concealed and in a dormant state —is not necessary to explain the fact of their having been here all the winter—if fact it be.

GROSBEAKS AND CROSSBILLS.

We have both these interesting birds in the Wye Valley, and though rare, for some unknown reason their numbers seem to be on the increase, more of them having been observed of late years than formerly. This winter the Grosbeak, or—as it is usually called—the Hawfinch

(*Loxia coccothraustes*), is quite common in the country around Ross. Captain Manly Power, of Hill Court, tells me he has noticed several of them upon the trees in his park, and the Rev. W. Tweed, of Bridstow, has also repeatedly seen them in his ornamental grounds, one specimen having been obtained and preserved by him. The severe weather may account for the numbers recently observed, in one of

CROSSBILL.

two ways—either that being a winter visitant its severity has sent more of them into our island, or the bird being shy—for it is one of the shyest of the Finch family—the hard times had tamed, and brought it down from the tops of high trees,—its usual perching-place,—and so closer to the observing eye. Though generally supposed to be migratory, there is reason to believe that a few pairs breed in this neighbourhood, and remain with us all the year. It is a bird well-known to the denizens of the Forest of Dean.

The Crossbill (*Loxia curvirostra*), a yet more interesting bird, is certainly a permanent resident in many parts of Herefordshire, as also becoming, year after year, more abundant. Mr. James W. Lloyd, of Kington, records it as occurring in that neighbourhood in the months of May, July, August, October, and during the winter; and since, in August, he has observed the male, female, and young together, it seems conclusive that they had nested there. The Rev. Clement Ley, of King's Caple, and Arthur Armitage, Esq., of Dadnor, have frequently observed small flocks of them, noting also that they usually appear in the mysterious odd numbers of three, five, or seven. In a very interesting article on the ornithology of Herefordshire, the joint production of these gentlemen, it is remarked: "Most curious birds they are, and very interesting it has been to watch their parrot-like motions, as they clamber from bough to bough of the spruce fir-trees, frequently breaking off a spray with the cone attached to it, which they grasp in their claws while they extract the seeds, producing a loud, snapping noise with their powerful bills. Among those which visited us last summer were several young males of the year, whose brilliant rosy plumage formed a striking contrast to the almost sooty hues of their companions." In fact, taking its habits, colour of plumage, and general appearance into account, the Crossbill is as much parrot as finch, and several of the species, as *Loxia pityopsitaccus*, *L. taenioptera*, and *L. leucoptera*—all occasional, though rarer, visitants to our island—show this alliance in an equal or even greater degree.

THE CARRION CROW.

The Carrion Crow, that is, the real crow (*Corvus corone*)—since the Rook (*C. frugilegus*) is sometimes so called—commences its nidification early in March, either repairing the old nest or building a new one. The Carrion Crow, however ill-sounding its name or wicked its propensities, has at least one virtue deserving a word in its favour—it is faithful in its loves. Even the dove, emblem of constancy, is not more true to its mate than this bird of reputation black as its plumage. And while the mated birds are constant as husband and wife, they are equally affectionate as father and mother; the young remaining under their protection, and possibly receiving instructions from them throughout the year, or until they get married themselves. The naturalist of Selborne, apparently quoting from Pennant's "British Zoology," says that "Crows go in pairs the whole year round." This is an error; they are only seen in pairs during the few weeks when engaged in bringing forth their young, after which they are rarely ever apart from these last. The family group usually numbers five or six, though often there are as many as seven. If the nest has been plundered—no rare occurrence when boys or gamekeepers are about—then may the two old birds be seen alone for the rest of the year. Notwithstanding its name, this bird does not confine itself to eating carrion, but often subsists on insects and reptiles taken alive. It will even kill young rabbits and leverets. It has also the repute of making free with the young fowls of the farm-yard; but I believe that much of this sort of damage laid to its charge is the doing of the magpie, which last sly depredator steals many a march into the outhouse enclosures, and carries off weakling chicks, despite the protecting efforts of the enraged parent. The Crow otherwise is not only innocuous, but of great benefit to the farmer, its principal food being the larvæ of noxious insects. It is especially destructive to the *scarabidæ*, and in search of these explores every dropping of cattle, often scattering the heaps, which, left untouched, would be injurious to the after pasture. In regard to these droppings, I have observed a fact worth recording. It is well known that cows will not eat the grass which grows out of their own ordure. I had a pasture field where this was plenteous, the rank spots showing conspicuous all over it, into which two of my horses were turned; and while the former carefully shunned the rich succulent herbage originating from themselves, the latter greedily ate it, browsing it down to the roots !

Reverting to the Carrion Crow, it takes a practised eye to tell one of these birds from a rook at 200 yards distance. There is scarce any appreciable difference in their size, shape, or colour, while they are almost as one in gait and general action. Seen near enough, however, there is no difficulty in distinguishing the species, the bare triangular disc at the base of the rook's bill being the best mark of distinction. Several pairs of Carrion Crows breed in Penyard Wood, each couple solitary, and not in companionship, as do the rooks. Just above my house, in the trees which grow against a steep escarpment, nests a pair, which I look upon as my especial pets. They spend most of their time on a stretch of pasture visible from my drawing-room windows, they and their last year's progeny stalking carelessly and majestically about among my black, but white-faced and white-tailed, sheep, already known to fame. Last summer, in the haymaking time, provoked by the loss of some chicks and ducklings, supposed to have been carried off by these crows, I was cruel enough to use my gun, and fired at one of them. Luckily I did not kill, but only wounded it in the leg. For many weeks after I saw this same bird limping about over my lawn ; and, at the time, a cripple myself, I could not help thinking it appeared there as a reproach to me, saying, "Just see what you have done ! Look at me, and then at yourself !" I was glad to find that its leg was not broken, and to see it recover, till at length it walked, and still walks, as well as any of the family. But the incident taught me a lesson of humanity, and never again shall my gun be discharged at Carrion or other crow.

THE NEST OF "CORVUS CORONE."

The nest of a Carrion Crow has been brought me for examination ; a nest which the owners had abandoned. Likely enough, its egg treasures had been taken out by some scansorial plunderer, as the eggs of this bird, being rather pretty and of large size, are a desideratum in collections. As many people suppose that the nest of the Crow is similar to that of the magpie, it may be worth while giving a detailed description of it, since, in many essential points, it differs from the latter. What may be termed the outer wall of a magpie's nest is composed of *dead* sticks, these nearly always branches of the hawthorn and blackthorn; some of them are thick as a finger, and so attached to the tree in which the nest is placed that the removal of it would either entail its destruction or cutting off the tree's top. With the Crow's nest it is different ; this being set in a fork of the trunk with little or no fastening, and can be lifted out bodily without breaking it up. Besides, the materials of the outer wall are not thorns, but the slender twigs of other trees, none of them thicker than a penholder. Those in the nest before me are nearly all oak, with a strand or two of honeysuckle entwined, evidently to bind them together. But what seems strangest about it is, that the oak twigs are all freshly torn from the tree or trees. I say *torn*, since each shows a ragged end, quite different from what would appear had it been snapped or broken off, and as if detached by a process of pulling and twisting. Now, as this nest was in an oak standing amidst other oaks in a wood, the twigs, no doubt, were obtained from the trees around, and, I believe, plucked from them by the birds themselves, since there are none lying loose upon the ground, and no work going on in the wood where sprays of this description could be obtained. There are nearly two hundred of these slender rods forming the outer wall, bent round it, and slightly wattled.

Again, a magpie's nest is usually domed over, while that of the Crow is quite open at the top, the whole structure being hemispherical. The one before me is eighteen inches in diameter across the top, of which the wall of twigs, with inside lining included, occupies one half, being about four and a half inches thick. The author of "The Gamekeeper at Home," speaking of Carrion Crows, says : " The keeper smites them hip and thigh, and if he comes across the nest placed on the broad top of a pollard tree—not on the branches, but on

the trunk—sends his shot through it, to smash the eggs."
I never heard of a crow's nest "on the broad top of a pollard
tree"; but whether there or elsewhere, I should say that
the keeper who acts as above were a man without much
intelligence, and silly in so wasting his ammunition.
For the wall-work of a Carrion Crow's nest is so thick,
and of such solid structure, no shot of gun, save the
bullet of a big bore, could possible be sent through it.

Perhaps the most notable difference in the nests of
these two birds is their lining, with the materials com-
posing it. In the magpie's nest there is only one layer,
which is some sort of threadlike, fibrous substance, apparently the root processes of the ivy. A compost of
mud, or clay where it can be had, is laid underneath
these rootlets to attach them to the wattle-work of sticks.
The interior of the Crow's nest is altogether different,
there being *two* layers of lining composed of various
materials. Nor is there any mud, clay, or such earthy
matters, though Montagu and other ornithologists
say there is. I myself have never seen such in the
Crows' nests that have come under my observation. The
lining, as I have said, is two-fold: first, a layer of grass,
this also of two sorts—*cooch*, and a broad-leafed kind
common in our woods, and known to the woodmen as
"*deer* grass." These mixed together form a stratum of
an inch thick, resting immediately on the groundwork of
twigs; while the extreme inner lining, of about the same
thickness, is composed of many substances, combined
and closely *felted* together so as to make a neat hemispherical cavity. Pulling them apart, I find horsehair to
predominate with wool; and a few birds' feathers, among
them two or three wing primaries of the wild pigeon
(Quest). But, *mirabile dictu!* human hair also, woman's

G

hair, a thick tress of it, full fourteen inches in length, and of a beautiful dark auburn hue.   Where this could have come from, or how the Crows got hold of it, it is hard to say.   Were it a short twisted tuft, one might believe it the castaway combings of a head; but a tress of such thickness, length, and beauty, where is the woman or girl likely to part with the precious treasure?

Nor is this all, my Crow's nest affording still other food for curious reflection.   In its varied material of lining are several scraps of old newspapers; which, as with the nest of a chaffinch I have already given account of, have been taken chiefly from the advertising columns, these two setting forth the merits of various patent medicines. Conspicuous among them is a "cure all," warranted to relieve every ill flesh is heir to.   I refrain from giving the name of this wonderful specific, lest I might be accused of puffing it.   Therefore the curious must be contented with my telling them it is *not* "Cockle's Antibilious Pills."

<hr />

THE CROW A FAMILY BIRD.

Acute and conscientious as was the naturalist of Selborne, he has made some mistakes; one about the Crow, or, as commonly called, "Carrion" Crow (*Corvus corone*), which he tells us "goes in pairs the whole year round." An error that, with many more in relation to the habits of this bird, has been perpetuated by Yarrell and most other English ornithologists, so as to become the stereotyped phraseology of the encyclopedias.

I am able to state for certain that the Crow never

goes *in pairs* save during the days of nest-building. If seen thus at any other period of the year, it is because the nest has been robbed, or the brood in some way destroyed, leaving the bereaved parent birds alone for the length of another twelvemonth. But when successful in the hatching and bringing up their young, there is no separation nor pairing. Instead, the whole family keeps together—though apart from all others—throughout after summer, autumn, and winter, on till nesting-time in spring.

To verify this habit, I have been for years observing the behaviour of the bird, and can now vouch for it as a fact. My opportunities are excellent, as the Carrion Crow is common in my neighbourhood, more than one family having their cantonments near. A pair annually breed in a hanging wood contiguous to my grounds, and last year they were successful in raising their brood of four; since which time all six—the old with the young—have consorted together, never for an hour being apart.

At the same time I know of a single pair, not far off, keeping by themselves. But I know also that this want of sociality is not their natural habit, but forced upon them, either by bird-nesting boy or the gun of the gamekeeper.

THE CARRION CROW A CLEANLY BIRD.

Notwithstanding the foul habits attributed to the Carrion Crow, even to giving it its common name, it is in *person* one of the cleanliest of birds, and addicted to frequent ablutions. Even in the cold days of winter I have often seen Carrion Crows washing themselves in a brook that runs through my land; and but three days ago I saw one on a spot of grass meadow which the brook had overflown, the bird plunging and rolling about in the water with apparent delight, while it sent the spray in showers all around it. After the bath it flew up to a tree near by, and there alighting, shook the water from its body and wings, then went on preening its feathers, at intervals giving them a fresh, vigorous shake. As there was but the one in sight, I take it for granted it was the cock bird, the hen being at the time on her nest. Had it been otherwise, the pair would certainly have been together, or in visible proximity, for, winter or summer, the *Corvus corone* is never seen alone, save when its mate and the younger members of its family have fallen victims to gin or gun.

ABRUPT DISAPPEARANCE OF CROWS AND MAGPIES.

While out on a long drive, I was once struck with the almost total disappearance of Crows and Magpies from places where previous to that day I had been accustomed to see them. Going the same round but a few days before, I had observed the latter in troops of ten, twenty, or thirty, loudly vociferous, their chattering scarce ever out of my ears. Now only one, or at most a pair, is to be seen at a time, and silent as mice. The explanation is, of course, that these birds have mated, and gone about building their nests, or repairing the old ones of last year. When so occupied, the Magpie is shy, if not sly ; and will accumulate a half-barrowful of faggots on the top of a Scotch fir close to your house, and, it may be, right over your head, without your having seen it carry a stick thither !

The disappearance of the Crow (I speak of *Corvus corone*) from its customary haunts, much more interested me. For I may almost claim individual acquaintance with every bird of this species belonging to the parish I reside in, with parts of others adjacent. I at least know every family, with the field, ay, almost the exact spot, where each could have been found any day throughout the past winter. Their absence from these places told me they too were occupied with the building of new nests, or renovating the old ones.

MAGPIES; OR, THE ENGLISH BIRDS OF PARADISE.

It is scarce necessary to say that Magpies are numerous in Siluria, as in most places where woods abound. Just now, however, and for the two months past, any one passing along our roads might imagine them in greater numbers than they really are. For one of their habits, hitherto not much observed, is to congregate in the early days of spring, and remain so for several weeks ; the purpose evidently courtship, and the choosing of partners for the nesting season to ensue. I have counted as many as twenty thus together ; and their excited manner, with much vociferation, would lead one to believe that this

MAGPIE.

was the business they were about. It may be that the
old pairs are constant to one another, for certainly the
same nest is used year after year, and most likely by the
same birds. If so, the clamorous congregation may have
for object only the mating and marriage of the young
ones; and the chattering, oft in tone of angry objurga-
tion, may be disputes between their parents as to fitness
and settlement.

It is said that in some parts of England the Magpie
has become quite a *rara avis*, having been persecuted
almost to extermination by both farmer and gamekeeper.
A pity this, for it is one of our most beautiful and
interesting birds, its presence a cheer and ornament to
field and tree. A neighbourhood, or homestead, would
not seem English without it. And if Magpies do, now
and then, pilfer from the partridge's or pheasant's nest,
and carry off chick or duckling, they make amends for
such damage by destroying an infinite number of
noxious creatures, far more harmful than themselves.

It appears to me that this beautiful bird is ornitho-
logically the northern representative of the famed Birds
of Paradise of the tropics, of a nearly allied family, if not,
indeed, the same. Its voice, habits, close-set, velvet-like
plumage, with changing metallic tints, and, above all, its
ample development of tail, point to it as being a so-called
Paradise bird—that special to the more temperate climes.
Last year, while taking the young out of a Magpie's nest
for purposes of examination, I was impressed with this
fact in observing the behaviour of one of the parent
birds. Flying excitedly from tree to tree, now and then
clinging to a branch in upright attitude, with body elongated, wings outspread in a tremulous motion, and the
long trowel-shaped tail, with side feathers graduated in

regular *échelon*, all the while giving utterance to wild, agonized cries—very screams—it presented a spectacle beautiful as touching. If I mistake not, Mr. Wallace, in his fine book about the Oriental Archipelago, describes the Birds of Paradise as behaving in a very similar manner.

ARE MAGPIES GREGARIOUS?

As a rule, Magpies are seen singly, or in pairs, and most people know of the superstitious feeling attached to their appearance, as thus formulated :

> One for sorrow,
> Two for joy,
> Three a marriage,
> Four a boy.

Were there any truth in these old saws, and it is wonderful how they are believed in, some neighbourhoods would show a preponderance of sorrow; while in others the wedding bells would be kept constantly ringing, and places become over-peopled. This last would surely be the case in some of the western shires where woods abound, and where four or more Magpies consorting together is quite a common sight. In my own neighbourhood, the southern part of Herefordshire, it is almost a rarity to see but one or two together; and last week sixteen of the noisy chatterers were counted close to my house consorting in a single flock. This, however, is a somewhat singular occurrence, and no doubt due to the abnormal mildness of the weather, the Magpies mistaking it for spring. When spring comes, there will be nothing strange in it, as then these birds congregate in large assemblages, often of twenty or more, for courtship and marriage; and when married, models of constancy they become.

But, apart from their association at pairing time, and in the fields, I have evidence, lately gained, of their tendency to gregariousness, which I believe to be their real habit when in sufficient numbers to indulge in it. Three weeks ago my gun-man, instructed to get me a pair for purposes of scientific examination, found nigh a score of them in the same night roosting-place—for the time was just before nightfall. Nor were they roosted on tall timber, but among young oaks not much larger than apple trees, with the trunks ivy-entwined, and last year's leaves still on. A copse it is, of about an acre in extent, standing solitary and apart, though between two extensive tracts of woodland, and scarce two hundred yards from the edge of either. Why this preference for the copse as a roosting-place, over the continuous woods, is of itself a singular circumstance, and one I am unable to explain. Whether a better shelter or not, the latter would certainly have been a safer one, notably in the present instance, since my man had no difficulty in bringing down a pair of the birds as they screamed and fluttered among the branches such a little way above his head.

The Magpie, which I believe to be the temperate-zone representative of the tropical Birds of Paradise, is possessed of a beauty little known and too little appreciated. Viewed from a distance, only black and white colours in severe contrast are distinguishable; but taken in hand, just after being killed or caught, its iridescent raiment shows tints in brightness rivalling the hues of the rainbow.

It will be a surprise to ornithologists when I prove—as I hope ere long to be able to do—that in England we have *two distinct species* of this very familiar bird!

MAGPIES IN A MADHOUSE.

I have received account of a singular incident, furnished me by the chaplain of a west-county lunatic asylum, in which five of these birds were kept as pets for the amusement of the patients. They had been in the establishment before the chaplain received his appointment to it; and one day, shortly after entering on his duties, he was out in the grounds along with several others, when a Magpie flew towards him, alighted on his shoulder, and commenced nibbling at his ear. Astonished, and somewhat annoyed, he brushed the bird off; only to have it return again, and recommence the pecking process, which gave him no pain, as the thing was done gently, and seemingly in play. Still it tickled, while further astonishing him; all the more after repeated drivings off and back-comings of the bird. Not till then became he aware of the cause of its persistency, this a strange one.

It appeared, as told him, that he bore a striking likeness to a former patient in the asylum, lately deceased, who had been a favourite of this Magpie, the bird being his especial pet, and that the man had taught it the manœuvre which, misled by the personal resemblance, it was now essaying to practise on himself !

THE NESTING OF ROOKS AND MAGPIES.

Speaking of nests, a comparison between those of the Rook and Magpie suggests itself. Though so much alike, as seen on the tops of tall trees, closer examination shows many points of difference. That of the Magpie inclines to an oval shape, and is usually domed or otherwise roofed over. It is also a more elaborate structure, if I may use the expression, with more "basket work" about it, and firmly attached to the tree-fork. A Rook's nest is negligently constructed, with the sticks laid loosely upon one another rather than wattled. The winter blasts afford evidence of this difference in construction, as regards permanency, the former defying them even in the most exposed situations, while the latter gives way to them in places comparatively sheltered. The Magpie builds a house it intends to inhabit year after year, during its season of incubation, and for that purpose will return to it if left unmolested; whereas the Rook, though coming back again to the same place and tree, seems not to regard the labour of re-building.

THE CUNNING OF ROOKS IN THEIR CHOICE OF NESTING-PLACES.

I have often noticed the preference of Rooks for build-ing about churches, as a proof of instinct, or, to call it by its proper name, an act of ratiocination, admonish-ing them that such places afford greater security. I might have added that the same process of reasoning also guides them to build in gentlemen's parks, and by grand mansions, knowing these to be, if not sacred as the precincts of the church, equally, or even more, safe from the intrusion of nest-robbing boys and the danger of ten-shilling licensed guns. I know of parks where Rooks have their nests on trees so low, and, to coin a word, so *climbable*, that a six-years-old urchin might easily ascend to and despoil them either of their egg treasures or chicks. But the proprietors will not allow this; and so year after year the birds come back with equal, and, it may be, increased, confidence.

THE ROOK IN A COURT OF LAW.

Many people desire to have a rookery in proximity to their houses; nor can there be any wonder at this. For, despite some disagreeables attendant, the cawing of these birds, so familiar as to seem the conversation of friends, with the opportunity of watching their many quaint ways and movements, is certainly worth something. Yet is there great difficulty, as all know who have tried it, in getting Rooks to breed upon trees not self-chosen; and various artifices have been resorted to as attractions. It is not of these I intend speaking now, but to relate an anecdote furnished me by my friend W. Baker, Esq., of Lincoln's Inn, showing the Rook in a court of law, into which it was unwittingly dragged, as many of the human kind often are unwillingly. The Probate Court it was, the episode occurring in Ireland—Tipperary, too—where resided Mr. C., an old gentleman of large estates and noted eccentricity in his habits. Having a small rookery by his house, and wanting to enlarge it, by way of encouraging more birds to build, he had bundles of sticks cut into convenient lengths and laid in litter all round the place; which the Rooks, as is their wont, made free use of. But another eccentricity of Mr. C., which in the end proved less innocuous, was a mania for making wills. Many made he, year after year; so many and so varied in their conditions, as also the beneficiaries they referred to, that when he at length took departure from the world the difficulty was to determine which will was the latest made and legally valid one. As the natural consequence, there was dispute between several claimants, resulting in an expensive lawsuit of long continuance, epitome of which I give in Mr. Baker's own words, quoted from a letter lately received from him.

" When Mr. C. died, there was a lawsuit about his estate. Lord Longford *v.* Purdon was the name of the action, and I think it was compromised last year. One of the pieces of evidence produced to show that he (Mr. C.) was of unsound mind, was the fact that he *assisted his Rooks to build their nests !* My attention had been called to the case in a marked way, owing to the fact that in one will (unfortunately not the right one) he had named some connections of mine as legatees."

Had Barham, while writing his "Ingoldsby Legends," but known of this Tipperary incident, he might have given it a place alongside the "Jackdaw of Rheims."

THE JAY A CARRION FEEDER.

While an incident chronicled below shows the alliance of the Jay to hawks and shrikes, it also partakes to some extent of a vulturine character. For not only does it eat fresh meat of all sorts and kinds—even greedily devouring fat bacon—but will not disdain that which is tainted. I could recount many instances of its feeding on carrion—dead sheep left lying neglected near a wood's edge, or unburied offal thrown out by a farmstead—sharing the repulsive banquet with rat, stoat, weazel, magpie, and tomtit, to say nothing of *Canis domesticus*. Odd enough that in his stroll through Savernake Forest, the same in which he was witness to the encounter spoken of—a reverend friend of mine and his companions came upon the body of a dead deer—a carcase fast hastening to putrefaction—with a Jay perched upon it, "stocking" away with all its might!

The scientific names given to bird, quadruped, reptile, or insect should, where possible, set forth some indication of its character and habits. Unfortunately, this golden rule is too often disregarded, the vanity of naturalists—especially they of the closet—leading them to bestow titles complimentary to friends and patrons, so making the nomenclature of zoology unintelligible as ludicrous.

As regards the Jay being called *Garrulus glandarius*, there is nothing of this, the name being more a misconception. For after the habits of the bird, as above described and attested to, who could think it appropriate?

THE JAY A CANNIBAL BIRD.

I have never known Jays so numerous in my neighbourhood (South Herefordshire) as they are at the present time, and have been for a year or two back. Throughout the past winter, and the autumn preceding, it has been a common sight to see flocks of from half a dozen to a score skirmishing about orchards, or along high hawthorn hedges, screeching as though they would split open their throats. This is evidence sufficient that exceptionally inclement and trying winters, which make havoc among many other species of birds, have done no hurt to them. On the contrary, as I am inclined to think, that extremely rigorous winters are rather in their favour, providing them with, as it were, a perpetual feast, and the food most to their liking, which, I believe, is *not* acorn, but *flesh*. During the long-continued snows of January, 1880, and 1881, there was scarce a hedgerow that had not fieldfares and redwings lying dead alongside it, killed, not by the cold, but hunger; since in both years preceding the wild berry crop had failed, and everything else eatable by these migratory birds was for weeks buried up beyond their reach. Many of our permanently resident kinds perished also, but certainly no Jays, these finding sport, or at least plenteous sustenance, in what was death to the others. Cannibal-like, more than once have I seen them gorging themselves on the flesh of fieldfares that had fallen victims to the snow, seeming to hold revel over the unnatural banquet.

"GARRULUS GLANDARIUS" A MISNOMER.

It seems to me that the ornithological name bestowed upon this bird is a misnomer, both generically and specifically. Many other birds are as noisy chatterers, some even more so,—the magpie, for instance, and parrots in their wild state,—while it is not specially a feeder upon acorns. Neither does it seem correctly classed in the family *Corvidæ*, in which most ornithologists place it; for, though the smallest member of this family inhabiting England, it is, in reality, more rapacious than any of them, the raven not excepted. Besides, if anatomical structure be reliable as a guide to habits, the denticulated upper mandible of the Jay's beak, with its sharp curving claws, points to relationship with the *Falconidæ* quite as much as with the *Corvidæ*. But there is another family with which it seems to have a still closer kinship—the shrikes (*Laniadæ*). There is a striking resemblance between it and the great grey shrike, or butcher-bird (*Lanius excubitor*), not only in the *dentition*, but in many of their ways and habits, both being murderous birds. For I have reason to know that during the winters of 1879-80 and 1880-81 the Jays did not always await their weak bird brethren succumbing to death from starvation, but in many cases forestalled it by killing them.

It seems even less known, if, indeed, ever suspected, that the Jay often deals death to quadrupeds as well as birds. Quadrupeds, too, of no diminutive size, or without the strength to defend themselves, such as mice. For it will kill young rabbits, and, what is more, the squirrel, a robust, active animal, of pugnacious, predatory habit, which even the stoat often finds a doughty adversary. A well-attested case of Jays attempting the life of a squirrel, which would have been successful but for outside interference, has just come under my notice, furnished by my friend, the Rev. Arthur Armitage, chaplain of the Military College at Oxford. With some companions, he was exploring Savernake Forest, near Marlborough, Wiltshire, when their attention was attracted to a pair of Jays excitedly fluttering about among the branches, and giving utterance to their well-known screech, in tone harsher and seemingly angrier than usual. Drawing up to them, it was seen that they were engaged in combat with a squirrel, repeatedly darting at and pecking it; the quadruped doing the best it could to defend itself. So earnestly were the birds occupied with their murderous design, that the tourists got quite close to them before being perceived. Then desisting, the birds flew off, while the squirrel, disabled, was easily caught. On examination, it was found that one of its eyes was already gone, pecked clean out of the socket; while other injuries showed where it had suffered from the beaks of the Jays, sharp and hard as steel. Unquestionably they would have killed it outright but for the accidental interruption.

H

A LIVING JAY WITH BOTH LEGS BROKEN AND THE SKULL CRUSHED IN.

I have just received the legs of a Jay—lately shot in some preserved woods between Ross and Ledbury—both broken but healed up again. The bird had evidently been caught in a gin-trap, from which it had been taken by the gamekeeper and cast down as dead. And, besides the broken legs, a portion of its skull had been crushed in, as if by the butt of the gamekeeper's gun or the heel of his boot. All this damage must have been done to it months before, and yet the creature still lived, and when shot was in good condition, flying about among the trees as if it had never received injury ! The stoat, taken some two years ago in Shropshire, which had been several times trapped, leaving it only one leg, might be quoted as a parallel case. But, no ; as regards tenacity of life, I can believe anything of a member of the family *Mustelidæ*, especially after seeing, as I last summer did, one of its smallest species, a weasel, do battle with a large sheep dog for nearly an hour, before it was finally conquered and killed !

THE WAYS OF THE DORMOUSE.

The account I have given below of the Dormouse, as to its extracting nut kernels, has been confirmed by so accurate an observer as Mr. Harrison Weir. He seems to think, however, that the kernel is loose in the shell, and the animal turns the nut about, so as to bring it in contact with the hole already drilled. This is not so. The kernel adheres to the shell, filling up all its interior, and is scraped off piecemeal, as I described it. After a night's feeding—for it is by night the Dormouse does most of its eating—several nuts will be left with the kernel but partly consumed, these to be cleaned out at the next meal. I have examined them thus in all stages, from the shell half-full to only a small morsel remaining at the bottom, and invariably to see the gouge-like track of the creature's teeth all over the rasped (not gnawed) surface, this itself being always eaten down level, with no inequalities left save the marks of the incisors. The only part of the performance I am unable to explain is, how the detached pieces are extracted from the shell. The hole is too small to admit even the animal's snout, save with closed jaws, and thus it could not possibly take the chips up in its teeth. Therefore they must be got out by one of two ways—either by being spitted on the sharp-pointed incisors of the lower jaw, or licked up by the tongue. The latter, I take it, will prove to be the solution of the enigma.

And first, another note in connection with the hazel nuts. These are often without any kernel, a circumstance the Dormouse is not aware of till it has penetrated through the shell, making a hole not much larger than the head of a pin. Then, with the tongue no doubt, discovering there is nothing eatable inside, it drops that nut, and tackles on to another. And, as further proof that the creature's instincts are not infallible, but, indeed, rather blind, I have known it return to the same empty shell, and open a hole at the opposite end, to meet with a like disappointment. Whether while drilling this second hole it remembered having made the first one, I cannot tell; but I should think not, and that the useless afterwork was as that of certain tropical insects boring hole after hole through a thin board, as a place of deposit for their eggs, each time to find themselves back into daylight on the reverse side. A like delusion has been frequently observed in the case of the sand, or bank, martin repeatedly tunnelling its way through a mud wall.

Small as is the Dormouse—its weight being under an ounce—it is a very glutton, three kept by me in the same cage consuming in a single night three full-sized chestnuts and fifteen hazel nuts, the shells excepted. In bulk this mass of food eaten by each must have been equal to its own body, even exceeding it. But their digestive powers are great, and of quick action, as shown by the quantity of droppings every day needing to be cleared out of the cage. No sort of vegetable food offered them will be refused; and though nuts are undoubtedly their preferred diet, they have also a relish for apples. The largest pippin dropped into a Dormouse's cage will soon disappear, skin, seeds, and all.

I have said that they do their eating chiefly by night, and it is during the night hours they are awake and active, sleeping most part of the day. If kept in a dark place, there will be some modification in this habit, though not much. About daybreak they invariably retire to the dormitory compartment of the cage, to issue forth from it at, or after, sunset. For in their wild state they are habitually nocturnal, one of the reasons why so little is seen or known of them. Tree climbers they are, too, as much as pine martens or squirrels, if not more, though generally contenting themselves with the ascent of hedge bushes, or hazels, to the branches of which I have seen them clinging in all attitudes, back downward as often as otherwise. The prehensile power of their claws is not only great but something inexplicable. While handling a Dormouse in a semi-somnolent state, it caught hold of my finger by the claws of one of its hind feet, and hanging from it, absolutely dangling down, went off into a sound sleep; so remaining till my patience became exhausted and I released the finger from its clutch. Had it been left to itself I have no doubt it would thus have slept its sleep out. The bat, with its hooked wing membrane, could not well do more. But the true sleeping attitude of the Dormouse is with snout and root of tail in juxtaposition, rolled up in spherical shape—though not so perfect a sphere as the clewed hedgehog—with the long, bushy, and distichous tail coiled spirally around. When in its winter, or hybernating sleep, the creature feels cold to the touch, and one unacquainted with this singular phenomenon would suppose it dead. Hold it for a time in the hand, however, and, so warmed, its beard bristles will be seen to move, the body rise and fall in gentle resperation, till at length it awakes, gradually unfolding itself as it becomes conscious of existence.

Zoological writers place the Dormouse in the list of hybernating animals, and all believe it eminently so, as may be deduced from the name bestowed upon it. It certainly does hybernate, though, I fancy, not to the extent generally supposed. I had one brought me in the middle of January by a hedger, who had taken it while "pleaching" a hedge, at the bottom of which he found it, *wide awake.* And like enough in mild winters these little creatures are often up and about in the night, when there is no eye to observe them. If not, then their habits undergo change in confinement, and when kept in a house.

In such situations they also sleep soundly and heavily, unlike the sleep of ordinary animals, but, as a rule, only in the day. And if near a fire, never for any great length of time, its duration seemingly dependent on the temperature around them. If cold, they slumber on; if warm, they will awake.

I have much more to say about this most interesting quadruped; but, as my note has already outrun the allotted space, I must leave it over till another opportunity.

A CAGED DORMOUSE.

On the 23rd of May a bark-stripper brought me a dormouse which he had captured in Penyard Wood, and *in* its nest. This was fixed high up in a bunch of broad-leaved grass, known to the woodmen as " deer-grass," and was composed partly of the grass blades and partly of leaves of trees. Though a nest of the present year, strange to say, the Dormouse was not a young one, instead an old male, and wide awake when taken. The stripper tells me he has never known of an old one thus caught in the nest. As the latter was rather open at the top, contrary to what is usual, it may have been unfinished, and the animal in the act of adding to it.

Placed in a common bird-cage, and food offered it— shelled hazel-nuts and canary-seed—it refused to eat while under observation, showing shy and frightened. At night, however, when left to itself, it consumed a portion of both the nuts and seeds. Next day other eatables were introduced into the cage—lettuce, sorrel, and groundsel—all of which it ate, apparently with a relish.

This evinces a fact, I believe, not hitherto noted, that green vegetables form part of the food of the Dormouse. By most naturalists it is described as subsisting on acorns, beech-mast, grain, haws, and hazel-nuts, especially the last, from which it has derived its specific name, *Avellanarius*—a mistake and misnomer, according to Bell, who says: "The name *Avellanarius* is not well chosen, as the principal food of the Dormouse does not certainly consist of the hazel-nut; indeed, I have never seen any that could gnaw through the shell of that nut when fully ripe and dry."

My own observations, made on the one before me, are so far confirmatory of this view. After allowing it to hunger for two days, with only unshelled hazel-nuts in the cage, it did not gnaw through any of these shells, though it had tried several, as could be seen by its tooth-marks here and there over them.

Still the specific name, which was bestowed by Linnæus, may not be so inappropriate, since there can be no doubt of the Dormouse feeding upon hazel-nuts, and being fond of them. But, I think, it can only get at their kernels when in the green state, before the shells become hardened. At that time it is often seen perched upon hazel trees far above the height of a man's head, and seemingly as much at home there as any bird.

In reality, this beautiful little quadruped is more squirrel than mouse, though naturalists regard it as a connecting link holding half-way relations between the two. Certainly it looks like a miniature squirrel, the rufous colour of its coat and busby herring-bone tail giving it this appearance, while removing it from that of the mouse family, often so repulsive. Its habits, moreover, liken it more to the *Sciuridæ* than the *Muridæ*. It is a dweller among trees, or rather bushes, its prehensile claws giving it the power to climb, even to cling, with hinder feet, and at ease. Besides, it makes its nest among branches, as squirrels do. As these, too, it lays up a hoard of food—usually haws, hazel-nuts, and beech-nuts—and eats them seated on its haunches, and held in its fore-paws as by hands. It hybernates, also, as the squirrels, lying torpid and clewed up in a ball throughout the winter months. Yet I have known it awake in winter, even when very cold. Once in January, a hedger, *pleaching* a hedge near my house, caught a Dormouse curled up in dry grass at the bottom. If sleeping, it awoke, and showed considerable activity, at short intervals repeating its querulous cry in tiny treble. Taken home to the house, and put into a box-cage, it remained awake and lively afterwards; no doubt, from the indoors warmth. In very mild days of winter the squirrel rouses itself, and roams abroad; and certainly the Dormouse does the same; but from its smaller size and nocturnal habits it is less liable to observation.

Though the *habitat* of the Dormouse is usually remote from the habitations of men, no animal is more easily made a pet of. With slight care and training it will become tame and familiar, even to letting it run about loose; a thing to be avoided, however, if there be felines in the neighbourhood. I knew of one that went regularly to bed with its owner, sleeping indifferently in a fold of the counterpane, between the sheets, or coiled up under the edge of the pillow.

As a pet, many people esteem the Dormouse so much that half a guinea is often given for one ungrudgingly. This tells of their scarcity, for in no part of the British Isles, that I know of, is it found in any great numbers.

Its breeding nest is a hollow ball—as those of the harvest and wood, or long-tailed, mouse—the entrance rarely visible. It is sometimes set in a thick thorn-hedge or coppice; but the favourite nesting-place of the Dormouse seems to be in a young beech with bushy top, on which the leaves stay all the winter through. I once saw a family of dormice thus domiciled, just after the young ones had got big enough to be abroad. The beech, a mere sapling, with stem not more than an inch in diameter, and clear of branches for a yard or so, gave me an excellent opportunity for observing the behaviour of the little quadrupeds. They seemed to play as lambs, some running down the stem on one side as others went up on the opposite; and this in continuance, like the revolving links of an " endless chain."

THE SQUIRREL.

It is hardly necessary to say that in a wooded district, as is the greater portion of the Wye Valley, Squirrels abound. I have them in a grove in my own grounds, while on Penyard's wooded hill, and throughout the adjacent Forest of Dean, they are common enough. Our English Squirrel (*Sciurus vulgaris*) is interesting for many reasons. Its beauty, both of form and colour, its wonderful agility, with its many pretty ways, make it one of the greatest ornaments of our woodland scenery. Besides, it is the only indigenous tree-climbing and tree-inhabiting quadruped we have in our islands. I will not here enter into lengthened particulars of its ordinary habits or life-history, which will be found in zoological works and encyclopædias. All know that it makes its nest in a forking branch, usually high up in the tree, building it very much after the manner of birds, the material being twigs deftly interwoven, lined with moss and leaves. In this it brings forth its young, often as many as five in a litter. These for months follow the mother about, as chickens do the hen that has hatched

SQUIRREL.

them. The little kittens keep company with their parents all through the winter, and until the warmer weather of spring causes a break-up of the family circle, the separation being due to that instinct which leads to the perpetuation of their species. The Squirrel, classed among hyberna-ting animals, is not wholly so. On a warm winter day it wakes up, and strays about in search of something to

eat. It usually knows where to find this, since it is one
of the prudent creatures which lay up a store against
times of scarcity. As the nuthatch, its favourite food is
the hazel-nut, though it is also given to eating grain,
beech-mast, and a variety of other vegetable substances.
Unfortunately for its character of innocence, it does not
confine itself to these, but has been known to rob birds'
nests, sucking the eggs, and devouring the callow young.
Insect larvæ—of less consequence as regards the injury
done—it also makes an occasional meal of, proving this
playful rodent, supposed to be so harmless, a very voracious creature. But it does damage of a different and
still more serious kind—this to vegetation itself. Among
the items of its diet are the seeds of coniferous trees, for
which it has a *penchant* almost equalling that for the
hazel-nuts. It skilfully extracts them, sitting upon its
haunches, holding the cone between its paws, and peeling
off the scales with its teeth. If it went no further than
eating the seeds, no one would object. But unfortunately
it does go further; and in early spring, when the fir cones
are all empty or rotted by winter rains, and the young
leaf-buds begin to show upon the trees, the Squirrel
makes sad havoc among these. Still another kind of
damage it does, hitherto unknown to me, and of which I
have just heard.

One of the woodwards of the Forest of Dean informs
me that in the larch plantations over which he had ward-
ship for some years past, he had now and then noticed
large branches, and even tops of the trees themselves,
broken off by the wind. Some of them were of large
size, thick as a man's thigh; and for long he could not
tell why Eolus was dealing such wholesale destruction,
for there were acres upon acres of the larch woods

strewed with the dead and broken branches. He learnt
at length, discovering the cause to be Squirrels! Their
mode of procedure was by peeling off the bark, not only
in isolated patches, but in broad rings all round the
branch or bole of the tree—their object, of course, being
to eat it—and thus naturally killing so much of the
branch as was above, which, after a time decaying, gave
way before the wind. From the manner in which my
informant speaks of the circumstance, I fancy that hence-
forth Squirrels will be scarce in the Forest of Dean,
especially in that portion of it committed to his care.

I have spoken of the colour of our English Squirrel,
pronouncing it pretty. In its summer coat it is so certainly : above, a beautiful chestnut-red, and below, white.
In winter the upper parts become greyish, and in
northern countries, like many other animals, often nearly
pure white. But here, in Herefordshire, I have to record
a very singular family, as regards colour, which was
found in the Forest of Dean, not far from that pic-
turesque spot well-known to Wye tourists as " Symond's
Yat." One of the woodwards, already spoken of, going
his rounds in that neighbourhood, observed a Squirrel of
the usual reddish colour, but having a *snow-white tail*.
His idea was, that it might be an old one, age having im-
parted to it the hoary distinction. He thought no more
about the thing till, several weeks afterwards, when,
passing by the same place, he saw what he supposed to
be the same squirrel, but not now alone ; instead, accompanied, or rather followed, by five lesser squirrels, its
kittens, all with white tails, miniature imitations of the
mother! This time, having his gun with him, he could
not resist shooting the parent, while the quintette of kits
scampered off into the underwood, where he lost sight of

them. The dam was sent to a Monmouth taxidermist, by whom it was skinned, stuffed, and mounted, and long afterwards shown by him among other noted curiosities.

THE "HUT" OF THE SQUIRREL.

Fortunately for the naturalist, as the lover of nature, not all the snares, gins, and ten-shilling licensed guns can either exterminate this interesting quadruped or apparently much reduce it in numbers. In most wooded districts, despite all persecution, it maintains its ground; and from correspondents in Ireland and Scotland I learn that in both these countries for some years past its numbers appear to be increasing, rather than diminishing. In the Forest of Dean and other woods in the Welsh bordering shires squirrels are plenteous—too abundant, say the proprietors of fir plantations, to which, it cannot be denied, these animals do considerable damage, especially to larches. In hard winters they strip the bark from the branches round and round; and though there be but a twig thus bared, of course all the spray that is above it perishes. In early spring also they gnaw off the young leaf-buds, so injuring the health and retarding the growth of the tree. During later spring and summer they are destructive to birds' eggs, but in autumn nuts and acorns furnish them with their preferred food; the latter supply the staple of it, and both are hoarded for a winter store. But this note is not meant to give a detailed account of the squirrel's habits, only of its nest, or "hut," one of which I have now before me, taken from a tree in the Forest of Dean. It is of rounded form, and roofed, with side entrance, its bulk being about that of a child's head ; and, as with the nests of most birds, it is double-walled, having an outside layer and lining. The former is composed of coarse moss, with an admixture of sheep's wool, and, more sparingly employed, broad strips of bark, the thin outer epidermis of young oaks. The material of the inner wall or lining is alto- gether different, consisting exclusively of the under or sap-bark of the oak, split into fibre-like strands, some of them fine as sewing-thread, their ragged ends and edges showing evidence that the splitting had been done by the animal's teeth. Now, to procure this material, and hackle it into the required condition, must have cost the squirrel, or pair of squirrels, a considerable amount of labour, independent of the task of construction. The question, therefore, suggests itself, why this extra toil when other substances, seemingly equally suitable, were to be had in plenty around ? Just by the tree from which this hut was taken there grow grasses of several species, some of them slender-bladed as the bark fibres used in its lining, and these could have been had with but the slightest exertion of strength or teeth. The employment of some preferred material in the construc- tion of their nests is one of the most singular habits, or instincts, of birds, and one still hidden among the *arcana* of nature. But not less singular that squirrels, also nestbuilders, should show a like instinct, for of other huts I have examined the materials were the same.

THE SQUIRREL A PEST IN FIR PLANTATIONS.

In early spring squirrels do considerable damage in the Forest of Dean by eating the bark and leaf buds of the young larches. Some days ago my gunman—who, by the kind permission of my friend Sir James Campbell, has free range of the Forest—brought home to me a batch of squirrels he had shot. All were in fine condition, quite fat, and unlike animals late aroused from the slumber of hybernation, which likely they had not much indulged in during the past mild winter. There were in all six of these squirrels, and they differed a good deal in size, as also in colour; some being of a much more vivid red, with the coats glossier, than others. On opening their stomachs I found them filled with a greenish substance, so comminuted as to be unrecognisable, though it looked like young larch leaves gnawed to a pulp. Mingled with it were soft masses of a yellowish white stuff, I took to be ants' eggs, also pulped.

The squirrel, notwithstanding its pretty playful habits and innocent look, is one of the most vicious of quadrupeds, as also the most courageous; especially the female when the mother of young. At this time, if the nest or "hut" be approached by any one climbing up to it, she will assail the intruder with all the fury of an enraged cat, and has been known under such circumstances to bite people severely. They are equally valiant when attacked by dog or other animal; and one of my ferret-keeping friends assures me, that a ferret has more difficulty in conquering a squirrel than a weasel, and far more than in killing the fiercest rat. The boys who live around the Forest of Dean often gang together, on Sundays or idle holidays, and go in chase of squirrels.

Not allowed to carry a gun into the enclosures, their weapons are usually stones and sticks. Their mode of proceeding is for one boy to "swarm" up the tree in which a squirrel is seen, force the animal off it to another, and so on till they get it into a tree standing well apart from the others. Driven out of this, its last stronghold as it were, it has no resource but to leap to the ground, where a "surround" of its pursuers has been previously arranged for cutting off its retreat, in which they are often successful.

Squirrels are sometimes snared, not by set snares, but a running noose of fine spring wire fastened on the tip of a pole, light and long as a fishing rod. This, cautiously and dexterously handled, is slipped over the squirrel's head, as it lies quiet along the limb of a tree; when, at length, taking the alarm, and attempting to scamper off, the animal finds itself fast in the wire, to be instantly jerked to earth.

———

THE WEASEL FAMILY.

Writing these notes, specially intended for the comprehension of those who have given but little attention to zoological studies, I may be pardoned for repeating, what every naturalist knows, that in the British Isles there are six native species of the *Mustelidæ*, or Weasel tribe, and one of doubtful foreign origin. The former, all wild, are the Weasel itself,—typical representative and smallest of the family, the Stoat, Polecat, Pine, and Stone Martens; with the Otter, differing in genus; the

latter a tame or domesticated species—the Ferret. Mr.
St. John, in his very interesting work, "Wild Sports and
Natural History of the Highlands," speaking of these animals, describes them so correctly and with such graphic
felicity, that I cannot resist quoting from him:—"The
blood-thirstiness and ferocity of all the Weasel tribe is
perfectly wonderful. . . . The more blood they spill,
the more they long for, and are not content till every
animal they can get at is slain. A she ferret with a litter

WEASEL.

of young ones contrived to get loose a few nights back,
and instinctively made her way to the hen-house, accompanied by her six kittens, who were not nearly half-
grown; indeed, their eyes were not quite open. Seven
hens, and a number of tame rabbits, were killed before
they were discovered; and every animal that she killed,
notwithstanding its weight and size, was dragged to the
hutch in which the ferrets were kept; and, as they could

not get their victims through the hole by which they had escaped themselves, a perfect heap of dead bodies was collected round their hutch. When I looked out of my window in the morning, I had the satisfaction of seeing four of the young ferrets, covered with blood, dragging a hen (which I had flattered myself was about to hatch a brood of young pheasants) across the yard, which was between the hen-house and where these ferrets were kept; the remainder of them were assisting the old one in slaughtering some white rabbits. Their eagerness to escape again and renew their bloody attack showed the excited state the little wretches were in from this their first essay in killing."

The present note refers only to the smallest of the tribe, the Weasel itself (*Mustela vulgaris*), and dwells but on one of its habits, often observed in Siluria. Last summer some mowers, cutting the grass of a meadow, were induced to suspend their scythe-strokes by hearing a sharp, plaintive cry, which they knew to proceed from a rabbit. The meadow, a small one, was surrounded by woods, out of which bolted a large rabbit, and shortly after a weasel in pursuit. Instead of seeking shelter, as would seem natural, the rabbit kept to the open where the grass had been cut. Nor did it continue in a rush of retreat, but every now and then made a stop, repeating its frightened cry. Very fear, perhaps the certain knowledge of its fate being sealed, seemed to enfeeble and render it irresolute. Still, the ruthless pursuer, like a diminutive sleuth-hound, kept after it, though not all the while visible to the mowers. Even in the mown sward its elongated vermiform body was at times out of sight, as it paused in crouching attitude between the successive shoots and zigzags of the chase. After making several short stops, the rabbit appeared quite overpowered with fear, and, as if deeming escape hopeless, at length came to a dead standstill, seemingly with no thought or effort to go farther. It even half squatted down, as if to make it more convenient for the cruel pursuer to mount upon and make prey of it. All which the Weasel did in an instant after, springing on the rabbit's shoulders, and laying itself along the neck, the latter, with a last agonized cry, but almost without a struggle, falling prostrate on the grass. There was nothing particularly strange in all this, a spectacle the men had frequently witnessed before. The unusual part of it came after, when they observed the Weasel in a few seconds' time forsake the quarry it had killed, and go streaking back into the wood, out of which, in less than a minute more, bolted another rabbit, pursued in the same way, overtaken, and killed. But this was not all, nor the half of it. For still another rabbit was run from among the trees into the meadow, to be served in a similar fashion, and another and another, till *six* dead bodies were upon the sward—all apparently the work of one and the same Weasel!

It is a curious fact, and I believe it to be a fact, that the rabbit, when pursued by stoat or Weasel, never takes to its burrow; yet when chased by dogs this is the first place it makes for. It seems instinctively to know that in its subterraneous abode, secure against every other, it has no security against those its natural and worst enemies, but would there be more at their mercy than anywhere else.

THE RABBIT AND WEASEL.

That an animal so large as a rabbit—combative, too, as proved by fierce conflicts with those of its own kind— should be so easily conquered and destroyed by such a diminutive creature as a weasel, seems one of nature's wonders. Yet it is an incident of every-day occurrence— so frequent, indeed, that, despite the solitude of the stage on which this tragedy of animal life is usually enacted, the spectacle is often witnessed. A young lady on a visit at my house some time ago, while out for a walk in Penyard Wood, heard a shrill scream, somewhat resembling the cry of a child; and hastening towards the spot from which it seemed to come, saw a rabbit wildly rushing about in the middle of the wood-road. It was acting as if demented, though only making efforts to escape from a stoat, which had attacked it, and was seen clinging along its neck, head to head. Vain they proved, for soon as sighted, and but a few seconds after uttering its distressed cry, the rabbit sank to the earth, dying almost instantly; while the weasel was driven off. But so far from being frightened, this fierce creature, a very monster in vicious appearance, in its retreat several times turned round, and raising itself erect on its hind legs, seemed to threaten the young lady herself. I have often noticed the stoat assume this defiant attitude when disturbed at its sanguinary game.

A still more curious incident of this kind lately came under the observation of one of my servants. Out in a field not far from the house he saw a rabbit being chased by a weasel. The chase ended almost immediately by the pursuer springing upon the shoulders of the pursued, giving it the fatal bite on the back of the head, and bringing it to the ground. As the quarry fell close to where he stood, he at once chased the weasel off and took possession of the rabbit, finding it, as he expected, in the last gasp of life. And while he still stood holding it in his hands, a second rabbit came running along the edge of the field, exposing itself to the view of the weasel, not yet out of sight. Quick as thought, the bloodthirsty creature, deprived of one prey, hastened to lay hold of another, regardless of the presence and near proximity of the man. And it succeeded; in another second's time being seen to leap up to the rabbit's neck, bite, and bring it to the ground just as before. And as before it was driven off from its prey, both rabbits being retained by my man-servant, and handed over to the cook.

It is no uncommon thing for rabbits to be found lying dead about the fields skirting a piece of the hilly woodland infested by weasels—killed by the latter, of course. When recently killed, or fresh enough to be used for food, the finder generally so utilizes them.

The mode of attack on the part of the predatory animal—which I believe to be chiefly, if not exclusively the stoat (*Mustela erminea*), and not the smaller or common weasel (*M. vulgaris*)—is to lie in wait, or stealthily approach the rabbit when the latter is browsing at a distance from its burrow; then, with a rush and a leap, launching itself upon the victim's neck, and laying itself head to head. There is not much to come after— no combat, no attempt at defence on the part of the poor creature so assailed, only a repetition of shrill cries, which end abruptly by its dropping to the earth, if not actually dead, so paralysed with fear as to cease struggling altogether. And the tragic scene itself lasts only a few seconds longer, the stoat or weasel, after a very short interval, being seen to separate from its victim, and to go off after other game. A superficial observer, now regarding the dead body of the rabbit, and not aware of what had preceded, might wonder what had killed it; for there is no mutilation, much less any portion of flesh removed. Closely scanned, however, a wound will be discovered in the creature's head, the puncture of a vein, so slight as to appear done by a sharp-pointed instrument. Yet through this the life-blood of the animal has been drawn and sucked out to its very source, the four-footed vampire contenting itself with the blood, and caring not for the flesh, which it leaves to other carnivora of less fastidious tastes.

Sometimes rabbits are found with the head eaten off, the body remaining unmutilated and untouched. A case of this kind came under my notice only the other day. I believe this is the work of cats, not weasels, the cat always eating the head of hare and rabbit first, as by preference.

OUR WEASELS AS A STAIR OF SIX STEPS.

In relation to our native *Mustelidæ*, two points which seem to have escaped notice may be worthy of it : first, that within the limited area of our islands there are no less than six distinct species, taking the Otter as one ; and second, that in size, or at least length, they should be nicely, almost exactly, graduated as the steps of a stair. An average-sized Otter, from tip of snout to that of tail, will measure above forty inches, the Pine Marten thirty inches, and the Stone or Beach Marten about twenty-five inches; while the three lower representative types of *Mustela*, Fitchet, Stoat, and Common Weasel, are of the respective lengths of twenty, fifteen, and ten inches. There is yet another odd circumstance connected with this graduation of length, each species doubling on the next above and below, so as to make certain lengths of the chain, as it were, duplicate. In other words, the male Weasel is often as large, or even larger, than the female Stoat; the male Stoat in turn running up to the dimension of the female Fitchet, with a like proportion throughout the series of six! The graduation, however, as observed in the three smaller species, is more strikingly curious from their closer resemblance to one another. I have a collection of these before me of every possible size, from the little she Weasel of less than nine inches long, to the he Polecat (Fitchet) of over twenty inches. But all with such family resemblance, alike vicious in look, that one might easily imagine them members of the same family, only of different ages.

THE WHITE STOATS.

In a number of the *Illustrated Sporting and Dramatic News* I gave an account of two White Stoats taken in the parish of Flaxley, Gloucestershire, near the Forest of Dean boundary. One of them, or rather its skin stuffed and mounted, is in my possession; and I find that in describing them as white all over, save the tips of the tail, I made a mistake—at least about this one. Hurried examination of the skin, when turned inside out, hindered me from noticing that the frontlet and crown of the head, with a portion of the nape of the neck, still preserve their normal hue—a light bay. The other specimen, however, is nearly as white as a true Arctic ermine. That it is the cold which causes this blanching is generally acknowledged among naturalists, as also that an extreme degree of it is necessary to produce the change. Hence was I puzzled at its having come about in a winter so mild as the past one had been all through. On reflection, however, I think it likely that these stoats turned white in one or other of the two preceding winters,—perhaps partially in both,—and during the summers intervening they had not gone back to the bay colour.

It is rare to meet with White Stoats so far south as Gloucestershire, though instances have occurred, some even in Cornwall ; and Mr. Bell, in his "History of British Quadrupeds," tells us of such in the classic region of Selborne itself.

A PROLIFIC WEASEL.

In *The Live Stock Journal* of August 18th, 1881, I made mention of a gill Ferret that had brought forth eleven young, and was successfully suckling them with but eight teats ! They were then about a month old, and the owner, a labouring man in my employ, brought them to me for examination. Mother and all were enclosed in a rough deal box, and on removal of the lid, a curious spectacle was presented. Twelve vermiform creatures, looking as much reptile as quadruped, all white, with a yellowish tinge, the dam only distinguishable from her progeny by having her eyes open, and being a little larger than they. Not much, however, as even then the young ones were well grown. To see them crawl about the box, climbing over and pushing under one another, writhing and wriggling in knots and bunches, while giving utterance to querulous yelps, as so many puppies, was a sight quaintly curious. To me, however, the most interesting part of it was the fact of there being eleven of them at a birth; the usual number rarely exceeding seven, and the highest I had before heard of nine. So I determined to keep a watch over this wonderful brood, and see what would come of it; as I could hardly believe it possible for a ferret mother, Weasel though she were, to suckle *eleven* young with only *eight* teats, and raise them to full ferrethood. She did it, however—nursed and reared the whole " kit " of them till full-grown, with not a weakling among them! They are distributed now, disposed of to different people; and the prolific dam is dead, though she died not from having been so productive. Her death was brought about by exposure to cold, after the young had been taken away from her. The owner still retains three of them—a " hob " and two " gills "—having sold the other eight for half-a-crown apiece. The "hob," now trained both to rabbiting and ratting, was brought to my barn some days ago to clear it of infesting rats, which he did in good style, killing several, among them an old buck, which measured eighteen inches from snout to tip of tail. But neither did the ferret come off unscathed, as evinced by several scratches on its muzzle, made by the teeth of *Mus rattus.*

The owner of this remarkable family of domesticated weasels has made known to me a fact I was not hitherto, aware of—that the eyes of young ferrets do not open till five and sometimes six weeks after their birth; and, moreover, that one eye often opens days before the other.

WEASEL, WILD DUCK, AND WATERHEN.

One of my friends, who has an artificial pond in his grounds, directly in front of the house, and within view of the windows, was witness not long since to a somewhat curious spectacle. In the pond was a pair of wild ducks, or rather a duck and drake—mallards—pinioned to prevent them flying away. Some moorhens, or more properly waterhens, were there also, regular denizens of the place. While watching them, my friend observed a vermiform quadruped come out from among the evergreens, and go skulking around the edge of the pond, now darting this way, now that. There could be no mistaking a weasel, which it was, nor its design—evidently to make a meal upon one of the waterhens. But the drake, perceiving it, as if taking up the cudgels on their account—though more likely on his own and that of the duck—rushed out of the water, and, with open beak, hissing the while, dashed right at the weasel, which instantly turned tail, and scuttled back into the bushes. The singular part of the affair was in a weasel, which very rarely exhibits fear of any other animal, and will even show fight to a dog, thus retreating from an assailant so little formidable as a drake. Possibly the loud hissing and wing-flapping of the latter had for the moment disconcerted it.

RABBIT, FERRET, AND BADGER.

A poacher of my acquaintance (I admit holding correspondence with the fraternity) tells me that when there
is snow on the ground his ferrets have more difficulty
in running the rabbits out of their holes. The reason he
assigns is, that the rabbits, knowing the snow to be outside, feel that they would have but a poor chance to

THE RABBIT.

escape through it if they bolted; and so keep to the
burrow as long as they possibly can.

From the same authority I have it, that if rabbits
chased by a ferret take refuge in the "holt" of a
badger—as in their hurried retreat they often do—the
ferret will refuse to follow them in, as it would into their
own burrows. Keen of scent, as are all the *Mustelidæ*,

it is by this admonished of danger inside, and knows perfectly well what the danger is : a carnivorous creature, with jaws and teeth capable of killing it by a single " scrunch."

ROMANCING ABOUT THE MOLE.

There is, perhaps, no quadruped, of like diminutive size, about which so much has been written as the Mole, the subject seeming to have had a fascination for mammalogists, as that of the cuckoo for ornithologists. Bell, in his " History of British Quadrupeds," the accredited standard work on this department of our native *fauna*, devotes twenty-six pages to it ; while dismissing the badger with nine, the fox with eight ; giving the wild cat only five ; and to both species of the marten—perhaps the most interesting animals of all—scant twelve between the two. Alike voluminous have been other writers treating of the Mole ; and, were all that has been said of it true, its story would well merit such enlargement of detail. Even what *is* true fairly deserves this ; but most of the truthful is that portion of its life's history and habits that remains untold ; while whole chapters of fiction about it have long passed current as fact.

One of the most notable of these misrepresentations is the tale of the mole's so-called " fortress " or castle, which has not only been described by authors, but delineated by artists, the picture of it to be seen in nearly every illustrated work on quadrupeds, encyclopædias among the number. In the latest edition of Bell are given both horizontal and vertical sections of this remarkable strong-hold, showing its central hollow globe—the reposing place, as said, of the animal—with two parallel circular galleries, above and below, between which there are five passages of communication, and then, leading off from the lower one, nine other subterranean tunnels, in almost

MOLE'S PALACE.

regular radiation as the spokes of a wheel. Yet, for this wonderfully curious and symmetrical structure, so minutely described and delineated—in all seriousness, too—I do not believe there was ever an original. On the contrary, it seems certain that the whole thing is the fanciful con-ception of a Frenchman, Henri le Court, endorsed by Geoffroy St. Hilaire. To get at the truth and root of the story, I may quote from the account of it given by Mr. Bell, which is as follows :—

" For these and many other interesting particulars in the life and habits of the Mole, we principally are indebted to the researches of Henri le Court, a person who, having held a lucrative situation about the Court at the epoch of the French Revolution, retired from the horrors of that fearful period into the country, and there devoted the remainder of his life to the study of the habits of the Mole, and of the most efficient means for its extirpation. His discoveries have been recorded by M. Cadet de Vaux, in a work published in the year 1803, and more briefly by Geoffroy St. Hilaire, in his ' Cours d'Histoire Naturelle des Mammifères.' This distinguished naturalist indeed visited Le Court for the purpose of ascertaining the truth and extent of these discoveries, and of enjoying the facility with which he was enabled by long habit to trace and to demonstrate the various labours of this object of his incessant research."

Le Court's research seems to have reached further into the *arcana* of Nature, or under the ground, than those of any one else. I can, at least, answer for my own, since having opened scores of moles' nests—for they do make a nest—I never came upon anything of the fortress kind, nor ever met I mole-catcher who had, and I have cross-questioned no end of *Talparii*. What *Talpa* really constructs for breeding-place, and not as a retreat for repose or security, is a heap, or " tump," which externally bears a rough resemblance to the so-called fortress ; but internally, or rather subterraneously, is altogether different. There is a nest in the centre, mostly composed of the dead leaves of trees, and placed nearly on a level with the surface of the surrounding *terrain;* while leading away

from it are three or four underground galleries, neither
at equal distances apart, nor always alike in number;
evidently made without any aim or design, save that of
convenient passage out to different parts of the mole's
"mooting" grounds, and return from them.

And why should this animal contrive a stronghold of
security, since it has such in all its "runs," both the main
permanent ones and those of a temporary kind made in
the pursuit of worms? Only one of its enemies, the
weasel, can make way along either; and this slender
vermiform creature could just as easily enter and assail it
within "the fortress." In fine, I hesitate not to say, that
this self-same fortress, though described as being underground, were better characterized by calling it a "castle
in the air." I may be wrong, however, and, if so, will
be glad to be set right by some one who has actually
entered the Mole's fortress.

Another erroneous belief about the Mole, and one of
more important bearing, though with less of the ludicrous
in it, is that this animal benefits the farmer in various ways,
but chiefly by destroying wire-worms, which it is said
to feed upon. In *The Field* newspaper, some long time
ago, there appeared a communicated article alleging this
to be a fact, and backing the allegement with a string
of details, which, as I could see, were drawn from imagination, just as Le Court's castle. Yet neither *The Field's*
natural history editor, nor any of its numerous correspondents, has, so far as I know, contradicted the erroneous
statement, though it is calculated to do harm to the
agriculturist, by making him tender as to *Talpa* and chary
of destroying it. I can contradict it, however, proving
the Mole a real pest, showing, by many experiments
actually made, that *it does not eat wire-worms*, and will

not touch them, its sole food, so far as I have been able to discover, being the ordinary ground or earth-worm (*Lumbricus terrestris*).

THE MOLE A "CONFERRING BENEFACTOR"!

I lately noticed an article on the Mole, casting ridicule on all who destroy this little quadruped, which the writer believes to be a heaven-born blessing, while the farmer regards it as a curse, or, at all events, a very troublesome pest. In this the tiller of the soil is right, for a pest *Talpa* is, greater than rat, and ten times greater than rook. The writer in question says: "The mole more than makes up for any damage it does by destroying wire-worms and other grubs that prey upon the wheat crops." This is no new theory, and at first thought may appear plausible enough. It is not substantiated, however, by accurate observation, for *the mole does not eat wire-worms* and other noxious insect larvæ, the innocent earth-worm (*Lumbricus terrestris*) being its natural and regular food. The stomachs of many which I have examined contained only the latter, cut into sections for the convenience of swallowing; and that this is the mole's preferred diet I can offer very direct proof. One year early in June I had the luck to catch one alive—no common occurrence—and having placed it in a large empty flower-pot, from which an aloe had been just removed, I proceeded to experiment on its food partialities. Wire-worms it nosed and passed by, as though its palate disdained them; but as soon as an earth-worm was thrown in to it, the long wriggler was seized upon and chawed up with a surprising rapidity.

For reasons, presently to be stated, I was not permitted to carry out the experiment to my satisfaction; but this has been done for me by Mr. Allen, the very intelligent bailiff of my friend Arthur Armitage, Esq., J.P., of Dadnor, Herefordshire. Having captured a live mole, Mr. Allen placed it, just as I had done, in an empty flower-pot, where for three weeks he kept it, repeatedly, indeed every day, offering it wire-worms in abundance. *It would not eat one of them;* yet the moment earth-worms were thrown into the pot it gobbled them up greedily. Before the end of the three weeks it had become so tame as to take the worms out of his hand !

He says, moreover, that the "runs" of the moles in a field of young wheat are the favourite resort of the wire-worms, which affect loose, open ground; and these will be found plenteous in the runs, but never in the stomach of the mole. Mr. Allen assures me that the moles, besides being noxious in other ways, do great damage to young clover sprouting from the reed. It is then a tender, delicate plant; and the mole loosening the earth around its roots, causes it to wither and die.

Further experimenting on the mole I had caught, while my servant was searching for earth-worms to offer it, under some artificial rockwork, a toad was turned up, which I directed also to be thrown into the flower-pot. Then I became witness of an episode somewhat singular. The mole, utterly regardless of my presence—or the string tied to one of its hind legs, with which I frequently pulled it about—at once made up to the toad, and caught the thigh of the latter between his teeth. It did not bite the batrachian, however; only seemed to play with it, or as if

K

speculating on what sort of a meal it would make. But
the toad was in a very agony of fear, as could be told by
its air and attitude.

Night coming on, I had several shovelsful of earth
thrown into the pot, and so left mole, toad, grubs, and
worms—the mole's leg released from its tether. In the
morning the quadruped was found dead on the surface of
the mould, while the batrachian, with some of the anne-
lids and larvæ, were alive underneath it.

Now, the question is, what killed the mole? It had
not been taken in a trap, or otherwise previously injured,
nor could it have died of hunger, as there were earth-
worms in plenty around it. Did a despairing sense of
captivity cause its death? If so, why did not the same
happen to the one with Mr. Allen, which lived three
weeks in captivity; indeed, until he released it? Having
become a sort of pet with him, he did not like killing it.

Then, was the death of my mole due to some venomous
substance exuded by the toad in the water-like fluid?
The last seems the most probable explanation.

The writer referred to above affirms that the mole "is
most assuredly a conferring benefactor on the farmer, and
by perforating the soil and throwing up earth it improves
the natural pastures." He seems to overlook the fact
that the mole's victim, the earth-worm, does all this in a
much better manner—so efficiently as to have had a
chapter devoted to it by England's earliest naturalist,
Gilbert White, and a whole book by her latest and greatest,
Charles Darwin. If the mole were such a benefactor to
the farmer, it is rather strange perverseness that he,
whether grazier or agriculturist, has been for hundreds of
years waging war upon it, many being annually out of
pocket considerable sums for its destruction. And money

well laid out; though the writer in question pronounces
it "ignorance," describing his own mole-catcher as star.
ing like a lunatic " when I told him rather than kill them
he would do me a favour if he would bring me a cartload
of live moles, and turn them down in my field." No
doubt the mole-catcher did stare, nor any wonder at his
doing so; for if among farmers a vote were taken as to
which of the two was the lunatic, the newspaper writer
would find himself hoisted high on his own petard.

THE "MOOTING" OF MOLES.

The lore relating to *talpa* seems absolutely inexhaust-
ible. A large volume might be written about the habits
of this underground animal, every page. telling of something strange. The more I observe, the greater grows
my record of facts relating to it, many of them to me
quite incomprehensible. As, for instance, I cannot
understand how the creature works its way through wet
earth—mud it may be—proceeding at a pace faster than
the most adroit spadesman can dig after it, and yet no
morsel of the soil adhering to its smooth, velvety coat,
which comes out of the ordeal clean as a new kid glove!
I am aware of the usual explanation put forth—about its
fur standing on end, or rather lying any way it may be
stroked. The premises are right enough, but the conclusion seems a lame one. Such may account for the
animal being able to go tail foremost along its subterranean galleries—as it does when these are too narrow for it

to turn in; but that its fur should remain clean because reversible, is quite a different matter, and, I think, calls for other explanation.   Doubtless, some secreted oleaginous substance protects it; but in what way is no less difficult to understand.

Again, how does the animal throw up its hills, or "tumps," as West country people call them?   In all that I have read of moles I find no explanation of this; indeed, no attempt at one; zoological writers seeming shy of alluding to the subject, doubtless from inability to deal with it.   Yet, in all the unexplained actions of animal life I know none more puzzling, and few more mechanically interesting.   Here we have a little round hole, less than two inches diameter, in the firm turf of a pasture field, through which a heap of underground earth, often a bushel measure of it, and at times even a barrowful, is thrown up in a single night.   Nor is this all of the night's doings; half a dozen, or it may be a dozen, smaller ones may be seen near by at varying distances, the work of one mole between the two daylights.

Still, it is not the wonderful exhibition of industry which perplexes; that were comprehensible enough. The puzzle is *how* the task is performed.   For the heap is in most cases a regular cone, so obtuse as to approach hemispherical shape, and there are no tracks nor sign to show that the mole comes out upon the surface while raising it; instead, every evidence it does not.   How, then, is the earth thrown up?   It cannot be by the animal's feet and claws, however well adapted these are —the fore ones especially—for burrowing.   There seems but one way possible : that the mould is projected upward by the creature's snout, just as is done by pigs when "rooting."   All the mole-catchers I have consulted agree in this being the *modus operandi,* and no doubt it is so, though still a puzzle.

There is yet another puzzle—as to whence comes the ejected earth. Some of it, of course, from the ground immediately underneath the vertical shaft; but it cannot all come thence. A bushel, or even half a bushel, of loose mould could not be got from a bulk of the firmest packed soil scarce so big as a man's fist; and there is no larger cavity just below the orifice. It must then be brought along the horizontal passages—the "runs." But how so? By pushing forward, or a series of backward scrapings? To these questions even the oldest *talparii* cannot give satisfactory answer. I have spoken of "tumps," where the tossed-up mould will quite fill a bushel; but there are exceptional ones of more than a barrowful. I have just measured one in a meadow near my house, of oblong form, to find the greater diameter 4ft., the lesser 3ft., and the vertical height or axis 18in! On scattering this vast heap, I discovered that no less than eleven "runs" led away from it, radiating to every side. Still there was no nest nor cavity within; though this might have been made later on, as the heap was but recently raised, and no doubt intended for the place of parturition.

MOLES IN THE MONTH OF MARCH.

"Mad as a March hare" has long been a proverbial expression, quite intelligible, though all may not understand it. Its origin has reference to the rutting season, when these animals, like most others, are seen running about in a state of unusual excitement. Just so is it with the mole in March, the period at which it gives way to instincts of propagation, and the time when the *talparii* reap their richest harvest. For now the males follow the females, and pursue one another, doing battle along the main runs, regardless of traps or other obstruction. While this state of things exists— which it does from the latter part of February till well on in March—the mole-catcher has a busy time of it, and the busier the better he likes it.

THE GARDEN MOLE.

Let not the reader imagine I am about to speak of a distinct species or variety of *talpa*, though the mole-catcher will tell you of a "garden mole." He means, however, only one found frequently in the garden, and for the trapping of which he will charge you double— that is, sixpence—instead of the ordinary price, which is threepence. His assigned reason for this, a valid one, is the greater difficulty of capturing the animal in garden ground, from the looseness of the mould and the greater irregularity of the "runs." Often days, or even weeks, may elapse before a pair of moles that have taken to the kitchen garden—where they do infinite damage· ·can be coaxed into the trap.

THE MOLE IN FULL "MOOT" AFTER EARTH-WORMS.

The main galleries, used in passing from place to place, are permanent, while the ordinary worm-runs are from time to time abandoned when the *annelidæ* have been all eaten out of them. Then the mole betakes itself to a fresh spot; and when it first breaks ground in this, a curious spectacle may be witnessed, should there be daylight to show it—that is, the frightened worms crawling out upon the surface and wriggling about, just as if the ground had been sluiced with salt water! They will be seen rising in front and to each side of the underground tunnel as it progresses, staying hidden among the grass till the destroyer has passed on.

There is no animal, however low in the scale of intelligence, but knows its natural enemy; and this behaviour of earth-worms is another proof, if any were needed, that they are the special prey and food of the mole.

THE MOLES OUT AND ABOUT.

I have elsewhere spoken of the mole no longer throwing up its "tumps," or hills; giving reason therefor—that the earth-worms are now every night above ground, so that *talpa* has no need to burrow after them beneath it. On April 14th, strolling about my fields, I noticed here and there a round hole, the orifice of a tunnel, which, on being probed with my cane, was found to descend some five or six inches vertically, then angle off horizontally. The inside of the cavity, cylindrical, and exactly two inches in diameter, was smooth as that of a new-laid drain-pipe. Of course, I knew it to be a "mole run," though not of the ordinary kind; instead, an upward shaft, made by the animal for nocturnal excursions over the pastures around. I have heard and read of the mole concealing this door of outcoming and ingoing, with earth heaped over, fearing betrayal by it. There is no truth in the statement; there is nothing around the circular orifice of the cavity, which is clean cut as the entrance to a sand swallow's nest. Throughout the winter such holes are never seen; for then the mole has no business above ground. It comes to the surface in winter too, but not every night, or with such frequency as to make a beaten path like that above described.

These summer holes of exit and entrance, well known to the mole-catcher, receive attention from him. They may be old, and out of use, or moles may be passing out and into them every night. For ascertaining the truth about this, and to save him the trouble of setting extra traps, a skilled *talparius* will lay two or three straws athwart the orifice, and await the result. If, after a time, the straws have been pushed out of place, or otherwise disturbed, then the inference is that a mole must have done it, and down goes a trap into the "run," the setter of it feeling pretty sure that on his next visit he will find the trigger sprung, and a dead mole squeezed flat between the iron grippers.

On 21st of December—shortest day in the year—my ploughman, while resting his team on turning at the headland, saw two moles issue out of the bank close by, one evidently pursuing the other, as shown by their excited manner. Soon as in the open, the pursued turned upon its pursuer in fierce, angry fight. Their mode of mutual assault, as described by the witness—a reliable one—was more like that of pigs than anything he could think of, repeatedly thrusting their snouts underneath, then with a hoist upward, each endeavouring to throw the other on its back. And, singularly enough, the noise they made—for they fought not in silence—also bore resemblance to the squeaking of young pigs, of course with a diminutive volume of sound.

Left to themselves, how long they would have battled, or in what way the combat might have ended, cannot be told. For it was brought to a termination by the plough-man himself killing both combatants on the spot, though not on the instant, as curiosity for a time restrained his destroying hand. Though living all his life in a district where moles abound, and spending most part of his time in the fields where they are at work, he had never before seen two of them together above ground, much less a pair so engaged. Indeed, to see a single mole on the surface—unless it be a dead one taken in a trap—is an uncommon sight; and the spectacle of a combat between them is so rare that one might live in the country all of a life—or for that matter fifty lives—without ever having an opportunity to witness it. I have never myself seen such, and, besides that related, have heard of but one other instance of it.

To the ordinary English labourer, the mole, or " hoont," as sometimes called, is a creature to be killed on sight, as rat, weasel, or snake; and, as soon as my man had satisfied his curiosity, he brought the combat to a close, with the lives of the combatants.

Even this was done in a somewhat original fashion.

Having nothing in his hands by way of weapon, and neither stick nor stone being near, he picked up an object which promised to serve his purpose—a ball of snow, for there was a slight covering of this on the ground, which one of the horses had cast from his hoof in turning. Flung at the fighting moles—still too earnest in their battle to note his approach—it knocked both over, killing one instantly and crippling the other, to be finished by a stamp with his heavy, hobnailed shoe.

Brought to me, I found, as might have been expected, that they were both males, though differing somewhat in size, as also in colour, the smaller one evidently a young "boar" of last season's littering, the other at least a year older. There was no wound or mark of teeth on either, a circumstance somewhat strange, as these carnivorous little quadrupeds are furnished with formidable incisors, and known to make terrible use of them in tearing up their prey. Might it be, that when thrown on its back the mole is helpless to right itself, as long-fleeced sheep and turtles, and knowing this the antagonist aims so as to capsize it ? The structure of the animal's body, with its short, inflexible legs, seems to point to such conclusion. Certainly moles are often found dead in the ditches, from no assignable cause, and never one with scar or scratch upon it that I have heard of. I have not made the experiment of placing them on their backs, but intend doing so with the first living " hoont" which falls into my hands.

THE "HOONT."

A striking feature of our fields just now, more especially the pastures, is the number of mole-hills, or, as
here called, "tumps," observable all over them. I never
beheld them in such profusion; on some meadows so thick
that there is almost as much of the surface covered with
these dark, circular heaps as with the grassy turf around
them; all recently thrown up too, or at least since the
commencement of the frosty weather. Some are so large
it seems almost incredible they could have been made by
a creature so diminutive as the mole, taking the time
into account; for one which I noticed in particular bore
resemblance to a barrowful of loose mould "dumped"
down on the grass, its freshness showing that it had been
the work of the night or day preceding. Not only itself,
but a row of others on each side indicating the "run,"
all equally recent, proclaimed the tunnelling to have been
done by this wonderful navvy within a period of twelve,
or, at most, twenty-four hours! And this when the
earth was frozen to a depth of several inches! For
during the December snow, which fell upon a frost
already gone deep into the ground, I saw many mole-
hills freshly thrown up. There is much in the natural
history and habits of this curious quadruped which needs
explanation. Even its mode of burrowing, if I mistake
not, has never been clearly comprehended. No more the
fact that, passing through what sort of soil it may—the
ferruginous earth of the red sandstone, or the white tilth
of the chalk formations, squeezing through ground wet
or dry—its soft, silky coat comes out unstained and un-
sullied, as if from a wrapping of tissue paper. I hope to
have an opportunity of returning to this subject, which

is of some scientific interest. Meanwhile I take leave of *talpa* by observing that here, in Herefordshire, it is rarely called by its proper name *mole*. "Hoont" is its designation among our rustics, while, by a strange perversity of nomenclature, its true title has been transferred to a different animal, the land *vole* (*Arvicola agrestis*). The error, no doubt, is due to the similitude of sound between "vole" and "mole."

CAN MOLES SEE?

"Blind as a mole" has long been a proverbial expression; like many other proverbs, untrue, because based on erroneous *data*. For not only can the *talpa* see, but it possesses powers of vision sufficiently acute for all the purposes of its semi-subterranean life.

Moles rarely come under close observation when living, being usually caught in spring traps, and so instantly killed. Then their eyes, becoming wholly or partially closed, can scarce be detected under the *ciliæ* of soft fur which forms a periphery around their sockets. With a live mole, such as some days ago I held in my hands, it is different; and I could see the little black orbs shining like jet, while made aware by the behaviour of the animal that they also saw me.

My *talparius* tells me that if he do not cover up his traps so as to exclude every ray of light from the runs in which they are set, the mole will not enter them. Seeing the suspicious framework of iron, with its smooth trigger plate, it will turn snout upward, "scrat" its way to the

surface, pass the trap, and dive back into the tunnel
beyond!

The belief in this animal's blindness, so common as to
be almost universal, is therefore groundless, though no
doubt it sees better in a dim light than in a bright one,
its habits, as with the bats and owls, being chiefly
nocturnal. But, if at all deficient in visual power, the

MOLE.

deficiency is fully compensated for by a high development in three of its other senses—smell, hearing, and
touch—and possibly the fourth, taste. Certainly it has
a discriminating palate, as I have proved by actual test ;
while it can hear acutely, the least noise in its neighbourhood causing it, if above ground, to plunge instantly
under, or suspend operations if excavating below. It is

for this reason moles are so seldom seen upon the surface, though they are oftener there than is supposed. The fall of a distant footstep is a signal for them to retreat to the covered gallery, which they will have reached ere the intruder is near enough to catch sight of them.

As highly developed, if not more so, is their sense of smell. Traps that have been too much handled they back away from; and an accomplished *talparius* will replace the human scent by that of the mole itself, well rubbing the trap, before setting it, against the body of one already caught.

My mole-catcher tells me of a still more effective method for deceiving *talpa* by the scent: making this attractive, instead of repellant. It is done by bottling up a number of earthworms, and so keeping them till they become fluxed into a jelly. A portion of this, aught but agreeable to human olfactories, dropped into the run near where the trap is set, will attract moles from near, and afar, as valerian would a cat. Though unable to be his voucher for this curious circumstance, I believe it to be a fact, knowing the man's truthfulness, with the absence of motive for misleading me.

As to the sense of touch, the mole evidently possesses that in a high degree, its long tapering muzzle, as the elephant's trunk, and the bill of snipe or woodcock, being furnished with nerves of great sensibility, enabling it to tell by the *feel* what it comes in contact with.

Take it all in all, this humble quadruped, supposed to be blind, and helpless beyond the common, is better furnished, both for attack and defence, than many others seemingly its superiors in sense capability.

———

A PAIR OF POSSIBLE WILD CATS, AND A PROBABLE THIRD.

About ten years ago some boys of this neighbourhood, while birdnesting in the Chase wood, two miles from the town of Ross, came across and killed what they supposed to be a wild cat. By the description of it which has been given me, it must have been either a real wild cat, or a *Felis domesticus* run wild; but if the latter, it was certainly one of an uncommon kind. The boys had two dogs with them, and their attention was attracted to the feline by seeing the dogs excitedly take stand, and begin barking by the mouth of a largish hole in the mound-like fence which encloses the woodland. A stick being thrust into the cavity, there came out only sounds—a spitting and "swearing," as my informants put it; but the punching, persisted in, brought forth a shaggy, savage-looking quadruped, which they took to be a cat of some kind. At first issuing from the hole, she made a spring at the boy who was nearest; but missing him, was tackled by the dogs. Neither of these, however, was of much mettle—one being a superannuated spaniel, the other a worthless cur; and left to themselves the cat could easily have conquered both, and would, so the boys, who are now grown men, have assured me. But these taking part in the scrimmage with sticks and stones, the wild grimalkin, over-matched, gave way, and retreated up a tree—the nearest to the spot. Unfortunately for her, it was a young oak of no great size or height, and the boys continuing to shower stones at her with all their strength, hitting her some hard blows, she bounded down again, and back into the hole. From this she was once more "prodded" out, and as before made a spring at the

nearest assailant, who chanced to be the biggest of the
boys and best armed—his weapon a heavy hedge-stake.
A well-aimed down-blow on the cat's skull stunned her,
and before she could recover from it, other blows
administered in quick repetition laid her out apparently
lifeless.   The boys believing her so, and caring more for
birds' nests than cat of any kind—little did they suspect
what an interesting animal they had dealings with!—were
about moving away from the ground, in fact, had gone
some distance, when, chancing to look back, they saw the
cat, they had supposed dead, tumbling about and making
futile attempts to get upon her feet—in which, left to
herself, she would doubtless have succeeded.   But returning they made a sure finish of her, cutting off the
tail, which alone they took away with them as a trophy.
What a pity it was not the whole pelt!   Had it been so,
this article would, no doubt, be more interesting, or, at
all events, less conjectural.

That tail is now lying on the table before me, with a
full note account of the episode—when it was taken, and
a description of the cat herself, correct and near as
remembered.   She was larger than the largest house
" Tom," more stoutly built, with a square head, and huge
broad paws, her coat shaggy, the colour a barred black
and grey, with a yellowish tinge—all characteristics of
the true wild cat.   But the tail before me, which I can
see for myself, this puzzles me.   Its colour is ringed
black and grey, so corresponding with that of the wild
species.   But then the hair on it, though coarse, is short,
the shape tapering, and its length—for I have the whole
of it from root to tip, or rather its skin—is less than six
inches, while that of a true wild cat is, or ought to be,
over eleven.   But no better does this short tail tally with

that of the *Felis domesticus,* except in the "taper."
Indeed, in length it is farther removed from the latter;
hence a mystery which baffles all my attempts to solve it.
The behaviour of the animal certainly seems to point to
its having been a real wild cat, for it repeatedly sprang
at its assailants, growling and "swearing" all the while.
And something more remains to be told. A bailiff who
had charge of the wood was met by the boys shortly
after, to whom they reported the encounter, with its
result; to be told by him, how glad he was they had
killed "that wild cat," and he wished they had "done
the same wi' t'other un." For it appears there were two
in the wood—likely male and female—the man adding
that they had been there some time, had done no end of
damage, destroying hares, rabbits, and pheasants, and
that he could never get near enough to shoot them, they
were so "wonderful cunnin' an' shy."

Unfortunately, this woodman has long ago gone to join
"the majority," else I should belikes know more about
the animal whose caudal appendage perplexes me.

I have received account of still another remarkable
feline killed at a later date—seven years ago—in the
parish of Foy, some six miles farther up the Wye. A
farmer, whose land lies along the river, destroyed this
one; and, like the birdnesting boys, thinking it of no
more value than rat or weasel, forthwith had it interred
—skin, tail, and all! Had he but known that I would
have given guineas for the skin, it would now, no doubt,
be among my mounted specimens, instead of gone to
decay under a muck heap. For the description I have
had of the animal—size, shape, colour, everything, this
time including the tail—seems conclusive evidence of its
having been a true *Felis catus.* The account of its doings

L

is also confirmatory of this view. It had not only killed and carried off several of the farmer's fowls and ducks, but those of others in the neighbourhood, besides destroying some tame cats, and badly maiming others, that had chanced to come in its way.

The farm in question is on the skirts of a wood of considerable extent—that of Perrystone—in which most likely the animal had its lair, issuing forth only for nocturnal forays. A ' hanging " wood it is, on a steep slope overlooking the river, in places almost precipitous, and the likeliest of " lays " for such a creature. Still it could not have been haunting there for any great length of time, with a gamekeeper all the while on the look-out for "vermin." Besides, it must have been a poacher of a most redoubtable kind. The probability then of its having been a real wild cat rests on the supposition of its having found its way thither from the Welsh mountains, following the course of the stream downward, perhaps here and there making temporary sojourn. And the same may be said of the one killed, and the other seen on Chasewood Hill, which also overlooks the river. Such a migration were not only possible, but probable enough ; since, among the wooded " dingles " where the Wye has some of its sources—very fastnesses—this now rare animal is believed still to have existence.

TAME CATS TURNING WILD.

The common house cat taking to the woods, and there remaining—in short, becoming, to all intents and purposes, a wild cat—is an occurrence by no means

rare in the valley of the Wye. A case has just come
under my observation of one thus voluntarily abandoning
house and home for a permanent residence *sub jove*, and
a life, if not merrier, more congenial to its feline nature,
under the greenwood tree. The animal in question—a
male, by the way—belonged to a near neighbour, whose
house stands contiguous to the borders of Penyard Wood;
and it was to this last that Tom betook himself. For
a time after his being missed it was supposed he had got
caught in a trap, or shot by some keeper. After awhile,
however, he was seen wandering through the wood, or
rather skulking about, his movements showing no sign
that he considered himself strayed or lost. Instead, he
appeared as much at home among the trees as though he
had never been outside standing timber, and all attempts
to capture, with a view of returning him to his owner,
were foiled by his immediate flight and retreat to the
most inaccessible fastnesses of the wood. He had, in
fact, become wild as its wildest denizen, and as shy of
man's presence as either badger or fox. For four years
he continued to live this free forest life, and doubtless
would have done so to the end of his days—indeed, did
till their end—which was a tragical one, as his life
terminated by his getting caught in a trap that had been
set for "vermin" of a very different kind. So fierce and
full of fight was he when approached in the trap, that it
was found necessary to kill him ere he could be released.

A circumstance connected with this incident is worthy
of consideration by the naturalist. In its wild condition
the animal had undergone a physical change quite as
great as that which had come over it morally. It had
grown more than double its former size in the domestic
state, thus contradicting the usually accepted doctrine

that animals are improved, or made bigger, by being brought under the dominion of man. Its coat, moreover, after the four years of freedom from restraint, was of the sleekest and glossiest, its whole appearance proving it in perfect health, with all the litheness and vigour of its feline kindred—the leopards, panthers, and tigers. This fact of a tame cat increasing in size when it turns wild has been often observed, and would seem to strengthen the argument of a descent from our indigenous wild species, now nearly extinct; the latter, as is well known, being much larger than the former. Whatever the truth of this matter, it is certain that tame cats always evince a tendency to take to the fields, and still more the woods, where these are near at hand, to stay in them for periods longer or shorter, in proportion as they there find suitable provender; and, furthermore, that cats noted for this sort of absenteeism are always those of greatest size and strength. The distance these straying grimalkins will wander from their own homes is something wonderful. One lately shot in Penyard Wood was identified by its very dissatisfied owner, who lives at a little clutch of houses called Crow Hill, quite three miles from the scene of the slaughter! Yet this cat was not "after kind," but skulking among the trees in quest of squirrels, rabbits, or leverets.

WILD RABBITS WONDERFULLY PROLIFIC.

Whatever the fact elsewhere, in this neighbourhood the wild rabbit is prolific to an extraordinary degree.

An instance has come to my knowledge of no less than ten young being found in the same nest, all presumably the litter and progeny of a single pair. And when it is taken into account that during the spring and summer months these animals breed as often as house pigeons, that is, bring forth a fresh brood every five or six weeks, the increase in their numbers may be set down as something very surprising. Were they not kept under by the multitude of their enemies—both beasts and birds of the rapacious order—they would soon overrun any country which claims them as part of its fauna, and make havoc of all that appertains to the industry of farm and garden. By good luck they are fairly palatable as an article of food, which guards against their ever becoming a pest altogether unprofitable.

A HARE WITH TWO SETS OF SUCKLINGS.

The hare, though not so prolific as its near congener, the rabbit, is nevertheless known to bring forth several times during a single season; and sometimes in such quick succession that the young of one gestation are not quite cleared out of the way before the litter following claims the fostering attention of the teat. Of this fact an instance came under the observation of one of my friends but a few summers ago. He was seated by the side of a wood with a pasture field adjoining, quietly smoking his cigar, when his attention was attracted to a doe hare, which, running out some short distance into the pasture, was there joined by a brace of leverets—her own, of course—these setting to and applying themselves

industriously to her teats. After awhile the mother gave
them a signal to desist, by striking her forepaws with
quick repetition on the turf, the strokes causing a sound
loud enough to be audible to the ears of my friend at
thirty yards' distance. The command was evidently
understood by the youngsters, and instantly obeyed by
them, as shown by their separating from the mother's
side, hopping off, and disappearing among some long
grass that grew near. As soon as they had left her, the
dam turned back towards the wood, and making her way
through a hawthorn hedge, continued on to a clump
of gorse, just inside the edge of the timber. Entering
under this, she was lost to the view of the spectator, who,
all the while remaining motionless, and quietly smoking
his cigar, had been the unobserved observer of this little
drama on nature's stage. But there was yet another act,
or scene, in store for him, soon after witnessed within
the wood, and under cover of the gorse. Having risen
to his feet, and approached the place stealthily, and
without making the slightest noise, he there beheld the
same old hare in her nest, in the act of being suckled by
a second pair of leverets, the tiniest creatures that could
be of their kind, to all appearance only a few hours old!
There could be no doubt of their being brothers and
sisters—or, it might be, half-brothers and half-sisters—of
the pair that had received nourishment on the pasture-
ground outside.

A SUSPECTED "BARK-STRIPPER."

The "wood," or "long-tailed field," mouse (*Mus
sylvaticus*) is one of the hoarders, often laying up stores

of nuts, beech mast, and other specialities of its food, in surprising quantity. A curious instance of this habit has come under my notice, recalling the story of the " Maid and Magpie," as also Barham's " Jackdaw of Rheims." A wood-bailiff—who has charge of a shooting-box belonging to one of my friends, and situated on the wood's edge, not far from my house—had gathered about a half-bushel of hazel-nuts, and deposited them in a shed, as stock to be drawn upon when desirable. They were in a canvas bag, left with the mouth open, the owner deeming them safe, since the shed was inside an enclosed yard, and no one had access to it save a bark-stripper employed on the estate, and believed to be an honest man.

Some weeks elapsed before the bailiff went back after his nuts, wanting some to crack and eat. But, lo! the bag was nearly empty, only a few nuts being found in its bottom! Of course, the bark-stripper came in for a suspicion of pilfering, even to direct accusation of it; which he denied, stoutly asseverating his innocence. To be disbelieved, nevertheless; and for a time the man lived under a cloud : his character gone, and his situation endangered. He would, in fact, have been discharged but for the discovery of the real thief, fortunately found out in time; this was neither more nor less than a wood-mouse, or possibly a pair of them. It, or they, had carried off the nuts and hoarded them; the place of storage they had selected being, for quaint curiosity, on a par with all else relating to the incident. In a dark corner of the shed were three vessels, that had been there lying neglected for a length of time. One was a little wooden keg, or " bottle " so called, of gallon measure, in which labouring men carry to the field their drink for the day ; the second was a tin can ; and the third an earthenware jar, or "greybeard," both of like capacity as the keg. All three were found full of hazel-nuts, choke-full to their necks, with just enough knawing on the nuts to tell of their having been transported thither by mice. And *illus sylvaticus* proved to be the culprit, from evidence obtained afterwards ; so clearing the character of the wrongly-suspected "stripper."

THE LITTLE GREBE.

There are few birds more generally distributed over the globe than the Little Grebe (*Podiceps minor*). The multiplicity of its vernacular names, as " dipper," " di-dipper," " dabchick," " ducker," " loon," and the like, each having a local significance, points to a wide range throughout the British Isles ; indeed, it is found all over them, wherever there is lake, pond, or stream of sufficient depth to give it security by diving. Even in the pools alongside railways, formed by excavations, and others where brick-clay has been dug out, if of any considerable size, a pair of dabchicks will have their *habitat* and breeding place, sometimes sharing it with the more showy water-hen.

All over Europe this bird exists, as in most parts of Asia and Africa. In America, too, I have met with it on the ponds and streams of the Mississippi Valley, there leading the same solitary life as in England, swimming about, and at intervals turning its quaint somersaults as it goes under water, but never taking wing till absolutely forced to it.

In February, 1881, one of my friends was fortunate

enough to capture a dabchick alive. It was caught on the Wye River, near the town of Ross, and displayed a remarkable pugnacity, biting and scratching at the hand which held it, just as do the tomtits. Placed in a tub of water, it dived instantly, and swam round and round underneath, its mode of subaqueous progression, as my friend describes it, resembling that of the frog. Minnows, water beetles, and other insects dropped into the tub it refused to touch, though likely, had it been kept longer, the promptings of hunger would have caused it to act differently. On the second day of its captivity my friend restored it to freedom, letting it off on a large pond in the neighbourhood, when it went under the water like a shot, not coming up again till nearly a score yards off.

The rapidity with which the Little Grebe disappears beneath the surface is something remarkable, in this respect equalling any of its kindred. When a boy, my first gun was a flintlock,—percussion pieces being then rare; and right well do I remember that to kill a didipper, in clear daylight, it was necessary to blind the flash from the pan with a screen of paper, or a leafy branch.

Some English ornithologists speak of this bird as migratory—disappearing in the winter. This, however, must be taken as referring to lakes, ponds, and other stagnant waters, when frozen up. Then the dabchick must needs shift quarters—*nolens volens*. But when it has its haunt on the running river—unless this be also icebound—I believe it sticks to it throughout the entire year. Certainly, it is not a migratory species in the sense of periodical migration.

THE TREE SPARROW.

This bird is, in most districts, of sufficient rarity to make it interesting and its possession desirable. I have a specimen before me, just shot in my grounds, a cock in winter plumage, and for those who find a difficulty in distinguishing it from the house sparrow, to which it bears a remarkable resemblance, I offer some *indices* that may be relied on. The Tree Sparrow is smaller than its congener of the farmstead, besides being of neater shape, and trimmer in the arrangement of its plumage. Two whitish bars traverse its wings diagonally where the house sparrow has but one. But the best point of distinction, or that easiest to determine, will be found in the colour of the crown—this in the *passer montanus* being a fairly good chestnut, while in the *passer domesticus* it is bluish grey.

In habits they are altogether different, the former a shy bird, keeping afield, and, if I mistake not, only associating in families, save during severe weather. Then it sometimes approaches the homestead perforce, consorting with others of the *fringillidæ*, to which it generically belongs.

Likely enough the Tree Sparrow is oftener seen than recognised, its similarity to the gable-end chatterer making its identification very difficult indeed.

THE GROSBEAK IN GREATER NUMBERS THAN SUPPOSED.

The grosbeak, or, as more commonly called, hawfinch

(*Loxia coccothraustes*), though still a scarce bird in Great Britain, seems of late years becoming more plentiful. At least, so it would appear in the Welsh bordering counties, where not only do they show themselves in winter, but throughout all the year, breeding and successfully rearing their young. I have had ample evidence of this by having had their eggs brought me—which should not have been done—and seeing the birds themselves in all stages of feather change, from fledglings to the fullest plumage ever attained. A remarkable bird is the hawfinch, and a beautiful one too; though what most strikes the observer is its grand development of beak, alongside which that of the bullfinch is as a bodkin to a crowbar. Well does this justify its trivial name, grosbeak, as its specific appellation, *coccothraustes* (berry-breaker), for no shell or rind of berry could well resist its crushing power. Hawfinch is equally or even more appropriate, since the haw is certainly its preferred food; not so much the pulp of the fruit as the aromatic kernel inside the stone, which last it can crush between its mandibles as though it were but the thinnest of egg-shells.

During one winter my gunman shot for me two or three specimens, and could have obtained more had I wished, or allowed it. As the haw crop has been unusually abundant, this may account for the greater abundance of these fiuches; and likely enough in years when the former fails the birds will be absent too, going elsewhere.

One reason why the grosbeak is so little observed is its very shy habit, for it is among the shiest of the *Fringillidæ*. In summer the foliage conceals it, while in winter, with the trees stripped bare, it keeps among the higher branches, even the tops of the tallest, and at such a distance off is not easily identified; its size, of course, seeming less on the high perch, where, no doubt, it is oft mistaken for chaff, bull, or greenfinch.

That the grosbeak is often in greater numbers in a neighbourhood than is generally supposed some proof is afforded by an incident occurring to one of my lady friends, who takes interest in the habits of birds. During one of the long lying snows she was accustomed to feed a flock on the lawn, but a little way from the house windows, and one day, among the tits, sparrows, chaffinches, and buntings, appeared a bird larger than any, which she recognised as a hawfinch. It was shy at first, but grew bolder as time passed, and there was none to disturb the feeding of the flock. Next morning it brought another along with it, and on the following day two more, till at length five of these grand finches became recipients of her bounty. Yet this was in a district of country where the local ornithologists had even doubted the fact of a hawfinch having ever been seen!

THE NEST OF THE BOTTLE BIRD.

April is the nesting season of our permanently resident birds, and I cannot resist touching on the subject, with special reference to certain of their nests. Travellers in tropical countries, and people at home reading accounts of them, regard with wonder the nests of the so-called "weaver birds," "tailor birds," and others that show ingenuity of construction. Yet I doubt whether any of these give evidence of greater textile skill than that of our own best nest-builder, the long-tailed tit

(*Parus caudatus*), or " bottle bird," as some country people call it, from the shape and style of its nest. One I have just made note of, a nest of this year, during March, with all the eggs in it, which is an unquestionable curiosity, besides a beautiful specimen of bird architecture. Of a nearly regular ovoid shape, its longer axis is a little over six inches, the measurement crossways being four and a half. It is placed vertically on a wild rose-bush, in a hedgerow, the smaller end upwards, in which is the entrance hole, that barely admits the insertion of my forefinger. The bird itself passing in or out must needs have a squeeze for it, small though the creature be. The inside furniture is a thick lining of feathers, in which I identify those of the jay, with other wild species; while the main wall of the nest is composed of green moss and wool, firmly woven, or rather *felted* together, and supported in the rose-bush by several branches worked in with the material. The outside layer or surface is not the least curious thing connected with it : this an encrusting of small lichen scales, set all over it so thickly as almost to conceal the greenery of the moss, and give it a sheen of silver grey. And, as if to heighten the effect, here and there are larger and lighter coloured blotches of a thin substance, I at first took for bits of tissue paper, but which, on examination, proved to be the gossamer envelopes of some species of insect in the pupal state. Likely enough the tits had eaten the *pupæ* themselves out of their silken coats, before they were converted into nest ornamentation.

It has long been matter of speculative surprise that a bird with such lengthy development of tail should build a nest seemingly so ill-suited and inconvenient for its uses. The inside cavity, however, is ample, ovoid in form as the whole structure, with a depth of four inches and a width of three—so giving room enough for the bird to turn about, even both hen and cock sometimes occupying it at the same time, as seen in this one.

THE WHITETHROAT—ITS FLIGHT AND SONG.

At this period of the year (early May) the Whitethroat (*Sylvia cinerea*) may be frequently seen mounting up into the air in a nearly vertical line, singing as it goes. The flight, though direct, is not continuous, but in starts or by stages, much after the manner of the skylark. It does not, however, ascend so high as the lark, some ten or fifteen yards being the summit of its soaring ambition; on attaining which, it poises for a few seconds, then flies back to the bush, or hedgerow, from which it started. The naturalist of Selborne mentions the Whitethroat as one of the few birds that "sing as they fly," very correctly describing the manner "by odd jerks and gesticulations."

The song of the Whitethroat has a certain resemblance to the first few strains of that of the blackcap. The latter, however, is of longer continuance, and the notes that succeed, the "inward melody" and "gentle modulations" spoken of by Gilbert White, are wanting to the former. Listening to this portion of the blackcap's lay, one might fancy it to proceed from the throat of a blackbird, singing in the heart of a grove or wood, at a far distance off.

A LILIPUTIAN COMBAT.

That the passions of hostility and anger are not confined to large animals, but felt with equal intensity by
the smallest, I had this day (May 10th) evidence, and of
a somewhat curious kind. The day being remarkably
fine and warm—indeed, hot for the month of May—I
had my wolf-skin robe carried out and spread under a
tree to recline upon. Lying along it, and listening to
the songs of birds—now so varied—observing also the
movements of many species of insects, which the hot sun
had stirred into activity, my attention was attracted to
one of the latter, in a larval state, by its odd movements.
It was making way over the smooth surface of a velvet-
covered cushion, but for which it is not likely I should
have noticed it; the creature at full stretch being little
over the fifth of an inch in length, and not the eighth in
thickness. It was white too, or cream-coloured, the
velvet being dark blue, so rendering it conspicuous by
the contrast. Its close proximity to my eye, and curious
mode of progression, led me to taking special notice of it;
the latter being made by repeated contractions and exten-
sions of the body, at each the creature rising and standing erect on one end, then pitching forward to its full
length, and with a jerk drawing the tail instantly after.
The same singular procedure I had often observed in
larvæ of a larger kind, as no doubt has every one else.
But though odd enough in these, it seemed still more so
in the little midget—certainly not bigger than a cheese-
mite—that was journeying across the cushion. I was
about taking my eye off it, when I saw coming in the
opposite direction another insect, of about the same size,
but perfect, not larval. A wingless crawler this was, but

with my limited knowledge of entomology I was unable
to identify either.   The new-comer was also light-
coloured; and the two proceeding in opposite directions,
but along the exact same line, it was evident they must
meet head to head.   This in point of fact they soon after
did, their heads coming in collision, both evidently
taken by surprise at the unexpected encounter.   But
instantly recovering from it, they began a battle of the
fiercest.   Though matched in size, the grub appeared to
be the more powerful and attacking party, its quick,
violent contortions seemingly meant to enfold and crush
the perfect insect, its adversary, while the latter looked
as though struggling to escape.   At that moment I
would have given five shillings for five minutes' use of a
microscope; as a glance through one would no doubt
have revealed the varying attitudes of these miniature
combatants, engaged in a strife, no doubt deadly as
between lion and tiger, if on a smaller scale.   I at first
supposed that the crawling insect was but endeavouring
to get out of the clutches of the jumping one, and expected soon to see it dead and devoured.   Not so, however, was the result ; for, after more than a minute spent
in wriggling and wrestling, the antagonist somehow or
other got separated, and the crawler crawled away,
apparently unharmed.   Then the behaviour of the grub
afforded me another spectacle, interesting as that which
had preceded, and further proving it the aggressor.   It
turned to and fro on the velvet, darting out its head, first
to one side then to the other, in rapid succession, as a
hound trying to recover a lost scent, evidently in search
of the escaped enemy !

Were our eyes magnifying glasses, in the world of
Liliputian life we should, no doubt, often witness hostile

encounters, with a display of passion fierce as that which rages in the breasts of bigger animals—even of man himself.

A DEVOURER OF FISH FRY.

The "dipper" is a great destroyer of little fish; and those engaged in pisciculture had need be on their guard against it. Proof of its voracious appetite has been lately furnished me by the behaviour of a pair of dabchicks that had their home on an artificial pond in the park of one of my friends living near. This pond, or lakelet, is fed by a running stream, and the owner wishing to stock it with trout, had some thousands of the fry of this fish put into it. For a time they seemed to do well; but then it was noticed that, day after day, they were decreasing in numbers, until at length only a few could be seen. At first there was some surprise at their disappearance, with mystery. But ere long the cause declared itself, on the dabchicks being watched in their diving; when it was discovered that each time one went under a young trout was brought up in its beak, and swallowed without ceremony. They had, in fact, been all along living on the fry as their almost exclusive diet. Æstheticism pleaded hard for retaining the dabchicks, as an ornament to the water, and on account of their quaint, curious ways. But more material tastes prevailed, to the destruction of the birds, for the preservation of the fish.

THE CIRL BUNTING.

I believe this remarkably handsome bird to be in much greater numbers all over the country than is generally supposed. The close resemblance it bears to the yellow-hammer, its near congener, no doubt leads to its being often mistaken for the latter. Indeed, to most country people it is unknown, just for this very reason; as the two are so like in size, colour, and general habits as to be indistinguishable from one another at any great distance. He would be sharp-eyed who could tell which from which at twenty yards off, unless a practised ornithologist. To him, however, there are distinctive marks by which either may be identified at a glance; and to enable an ordinary observer to do this, I may say that the best guide—or most conspicuous one—will be found on the throat; that of the Cirl Bunting being black, as though it were a black scarf, while the yellow-hammer is without this sombre distinction.

It is a somewhat curious coincidence that with several *genera* of our small birds there are two species of each usually found frequenting the same neighbourhood so like one another as to require close scrutiny for their identification. Notable examples are the two pipits (*Anthus arboreus* and *pratensis*), the common and tree sparrows (*Passer domesticus* and *montanus*); the pied and white wagtails (*Motacilla yarrellii* and *alba*) and the sky and wood larks (*Alauda arvensis* and *orborea*), the last pair, however, not so much alike as the others.

SINGULAR CAPTURE OF A WOODCOCK.

Some time ago a labouring man in my employ made capture of a woodcock under circumstances so peculiar that probably the like may never occur again. He was sauntering along one of the wood roads (Forest of Dean), the day being Sunday, when he saw a woodcock at some distance before him, close by the path's edge. On the ground, it was going at a run through the grass; as it had already sighted him, and was making off in retreat. His rapid advance upon it first brought it to the squat; then, as he drew nearer, and it saw no chance of concealing itself, the alternative of flight was determined upon. I believe it to be a fact that the woodcock in taking wing uses its bill to help it up into the air, by pressing the latter against the ground. Certainly before rising, as every sportsman may have observed, this bird is seen with head down and rump elevated, seemingly straining its neck, as if for a leverage to aid it upwards. Just so was this one doing when come upon by my labourer—a man who had seen the like before, being well acquainted with the woodcock and its ways. But now he saw what gave him a surprise, the bird convulsively fluttering its wings, as in a struggle, while, instead of flying away, it remained in the same spot, and so stayed till he got up, and laid hold of it. Then to find that it was already held in fast grip by the ground, into which it had dug its beak, and could not draw it out again! Strange as it may appear, I can vouch for this as an actual occurrence; though the only one of the kind I have ever heard, or am likely ever to hear of.

THE FAVOURITE FERRET WITH POACHERS.

In a confidential chat I had the other day with an old transgressor of the game laws, though no longer such, I am happy to say, he let me into a secret of the fraternity relating to ferrets. It appears that the white variety of these domesticated weasels is that preferred by the poacher, and for the following reasons : When "rabbiting" at night—the poacher's orthodox time—a white ferret is more easily seen, its colour making it conspicuous in the darkness, and its work, done in silence, with dodging into and out of the holes, can be more readily interpreted as affecting results. Besides, a white ferret is less liable to be lost than one of the dark or *fitchet* colour. The poacher knows that the latter variety is of a fiercer, gamer nature, hardier than the former, and better for work in the ordinary legitimate way. But for clandestine nocturnal duty he prefers the pink-eyed *albino ;* the reasons, as above, being good and substantial.

FERRETS AND THE WILD BIRDS' PROTECTION ACT.

I have lately come to know that the destruction of wild birds which accrues from the keeping of ferrets is something considerable. Not that the ferrets themselves are blamable in the matter, but their owners. In

almost every neighbourhood there are poor men, one or
more of them, quite apart from the fraternity of poachers,
who indulge in the luxury of keeping a ferret or two,
partly to make money by occasional rat-killing for the
farmer, and partly by the young ferrets—a numerous
progeny. As these animals do not live on air, but require
substantial food for their subsistence—a goodly amount
of it, too—their owner is often at a pinch for the providing
of it. The sheep's paunch, which costs him twopence, is
his best stand-by; but even this runs up to money, taking
into consideration his precarious wage of twelve shillings
a week, often reduced by days of rain or sickness. So, to
economize the expenditure on paunches, he has recourse
to the *feræ naturæ*, and of these the young of wild birds,
callow in their nests, are the easiest of procurement.
They are in this state, too, just at the time when the
young ferrets are querulously calling for food, and needing a large supply of it.

A tale of poacher cleverness, combined with audacity,
has been lately told me, the narrator vouching for its
truth. The hero of it, a noted transgressor of the game
laws, was out "rabbiting" on a certain moonlight night,
having with him a pair of ferrets, a dog, and the usual
paraphernalia of nets. The scene of his operations was
a warren by the wood's edge on the estate of a neighbouring gentleman, and several miles from the poacher's own
home. He had just entered the "weasels" when the
gentleman's gamekeeper dropped upon him, catching
him *in flagrante delicto.* Still, he found time, before the
keeper got forward, to pluck up his nets, clew them into
a ball, and fling them into some bushes near by. As a
right-of-way path ran past the place, and the man was
unknown to the keeper—with no other evidence of his

guilt apparent than the canine by his side—the latter, doubting the chances of a conviction, hesitated about taking him up; perhaps all the more from his being a stalwart, determined-looking fellow. But, just then, one of the ferrets—a white one—showed its snout at the entrance of the burrow; and down dropped the keeper on his knees to lay hold of it. The animal, however, which would have allowed its owner to catch it, when approached by the stranger shied back again into the hole out of sight. It was a wide-mouthed cavity against a sloping bank; so the keeper flung himself flat on his face, thrust head and shoulders in after, and commenced feeling for the ferret. Seeing him so situated, a grand strategic idea flashed across the brain of the poacher. He was standing by the side of a hedge lately "pleached," and plucking up one of the pointed stakes, he stuck it into the bank between the keeper's legs, close up to his hips, driving it in firm and fast. The man endeavouring to draw out again, and finding himself fixed, commenced a series of wrigglings, accompanied by angry objurgations, that seemed as if sent up from the bottom of a well. In time his voice changed to that of entreaty, begging to be released. He could not release himself, as the stake hindered him from backing out of the hole, and it was too high for him to lift his legs over it. But the poacher was pitiless, and gave no ear to his entreaties, alone busying himself about the recovery of his ferrets. These had, meanwhile, returned out of the burrow, and popping them into the ample pockets of his velveteen—his nets recovered, too—he ran away from the place, leaving the hapless keeper with his head in the rabbit-hole! And in this "fix" the man remained throughout the rest of the night, and till near noon of the next day; indeed, he might have died in it but for one of his watchers, who, chancing to come that way on his rounds, found and released him. For this sharp practice on the poacher's part no prosecution followed, nor was any action taken afterwards, which may be thought strange. But the gamekeeper was a new hand in the neighbourhood, besides the culprit being altogether unknown to him. And possibly he had no desire to identify him, not liking to make a noise about an affair in which he had himself cut such a ridiculous figure. Of course, the poacher kept it dark enough, and it is only known to the *initiated.*

POACHERS IN PETTICOATS.

A keeper employed in the Government Forest of Dean, whose enclosed boundary is but a few stones' throw from my house, tells me a sorrowful tale of his troubles with poachers. He says poaching by snare and trap is so rife throughout the Forest that he can hardly go his rounds, taking his dogs along, without one or other of the canines getting caught in a "gin"—the steel spring trap.

This I can credit, knowing how strong and numerous is the fraternity of poachers all around the Forest borders. Indeed, there are families in which this practice is hereditary, and has been followed for centuries—the descendants of those who stole the king's deer, when the antlered stag was among its denizens. And now that there are no deer in it to be stolen these thrifty people of the modern day have transferred their industry to the acquisition of hares, rabbits, and pheasants. But what rather amused me in the keeper's account of his miseries was to find that around the Forest there is not only a *fraternity* of poachers, but a *sisterhood* of them— in other words, women engaged in it as well as men. As a rule, the men are employed at other work, in the coalpits and iron-mines, so having scant time to look after traps. But their wives and daughters do this, some of them, as the keeper says, setting a wire snare, or planting a gin, with as much skill as could the men ; while, notwithstanding their impediment of loose drapery, they are equally quick and clever in getting out of his way, whenever he makes an attempt to come up with them.

A KINGFISHER KILLED BY A PERCH.

On a certain pond, some years ago, occurred another curious episode, not witnessed by any one, but made known by results. A kingfisher was found lying dead by its edge, the cause of death unmistakable : it had caught a perch, and tried to swallow it, but without success ; for the fish was still sticking in its throat, the spines having penetrated the bird's gullet, and so choked it.

Something more in connection with this unwitnessed spectacle of nature is worth noting. At the time it occurred the pond, a very small one, had been but a few days established, and perch put into it. The situation is far away from any other water in which there are fish, on high lying land, and the last place one might expect a kingfisher to be found in. A bird, too, of such rare occurrence anywhere. Yet this one, guided by some inexplicable instinct, or, more likely, a reasoning intelligence, had so soon discovered the remote speck of water, and to its misfortune the fish in it as well!

GOLDFINCHES FEEDING ON FIR CONES.

It is cold-blooded cruelty, absolutely bird murder, to use the gun upon a goldfinch; yet I, who say so, have been guilty of this very crime, and but a short while ago; not wantonly, the reader may well be assured, but in the cause of science, if that be any palliation. It was done to verify a fact lately communicated to me by a lady friend, and with which I had not previously been acquainted. She had seen a flock of goldfinches in a grove of Scotch firs pecking away at the cones. I suggested "crossbills," though doubtfully, knowing the lady possessed of ornithological knowledge, but knowing also that these birds with the beak awry are frequently observed in flocks among our fir plantations. But no; she was sure that those she had seen pecking at the cones were goldfinches. And so am I now, after the ruthless murder committed—a veritable thistle-feeder (*Fringilla carduelis*), shot down out of a Scotch fir, where it had been gorging, its crop found nearly full of seeds it had contrived to extract from the cones.

ERRONEOUS BELIEF ABOUT THE WILD BERRIES.

It is a common belief among unobservant people that if the crop of wild berries—haws, hips, and those of the holly—be unusually abundant a severe winter will follow. Nature, in her beneficence, say these people, so provides for her favourite creatures, the birds, not forgetting the beasts. But if nature be so benignant, why does she let either ever starve at all? As it is, they did—the birds any-how—in hundreds and thousands during the winters of 1879-80 and 1880-81. Both were severe enough to test the truth of the above belief; which they did, showing it not true at all, but absolutely erroneous. For, in both, not only were the wild berries unusually scarce, but in many districts altogether wanting. And as further proof con-firming the fallacy, the winter just passing away, mild throughout, has been one with the berry crop so plenteous as to redden hedge and bush everywhere—berries of all sorts—just when the birds could have well done without them !

Naturalists of an amiable disposition, but not always true to nature, are very fond of dwelling upon her be-nignance, some of them ever dinning it into our ears. How good and wise she is, say they, in her every act and design ! Wise she may be for purposes we know not of ; but as to her goodness, it would be difficult to con-ceive anything more apparently cruel than her whole scheme as regards the *feræ naturæ*, one species preying upon another, all over the earth, in an endless chain of hostility and destruction. The sad fact exists, and the purpose, though to us inscrutable, may be of the wisest and for the best—indeed, must be. But is this a reason why naturalists should stultify themselves by an over laudation of nature, telling her to her teeth she does that which certainly she does *not?*

GIPSIES AND HEDGEHOGS.

Around here we have both of these curious creatures in abundance : the biped attracted by the Forest of Dean and other Wyeside woods, where he is permitted free tenting-ground; the quadruped finding in the dry tussocky outskirts and underwood a *habitat* to its taste. Mention of the one almost invariably suggests thought of the other. For who has not heard of the gipsy's fondness for the urchin's flesh, and his original mode of cooking it—a bake in a ball of clay? But I have reason to doubt the correctness of what has been said about this culinary process. It is certainly not practised by any of the fraternity around here—indeed, not known to them. All with whom I have come in contact tell me that their mode of cooking the hedgehog is simply by roasting it on a stick, or other spit, over their ordinary "faggot fire," having first removed the skin and " offal "; the which, so far as this neighbourhood is concerned, does away with the pretty story of baking in a ball of clay.

There is no question, however, as to their partiality for the animal's flesh. Gipsies, young and old, are *friand* of the same, speak of it as a *bon-bouche*, and take much pains to procure it. In its capture they display wonderful skill and sagacity. Where an ordinary individual can perceive neither trace nor sign of hedgehog presence a gipsy will sight the creature's " spoor," and follow it up

to the den, unerringly as hound on the scent of hare.
Some are noted for superior cleverness in this speciality
of chase, and proud of it. In a camp I lately visited, a
gipsy woman—with, by the way, a very pretty daughter
at her side—while lauding the superiority of "hodgkins'"
flesh, also took occasion to sound the praises of her husband—who was absent—enthusiastically proclaiming him
a " good hedgehog dog,"—the best in the community.

A woodreeve of the Forest of Dean tells me that he
has seen as many as fifteen or sixteen hedgehogs in a
gipsy camp all at one time, hung up on the branches of
the trees, skinned, cleaned, and ready for the spit.

The ancient British Kingdom, or Principality of
"Ergyn," in Saxon and Norman times known as the
Hundred of Urchinfield, now Archenfield, whose territory
extended along the Wye from the Forest of Dean to
Hereford, no doubt drew its primitive name from being a
special abode of the urchin, *Britannicé,* " Ergyn." Possibly, too, the name of the Roman station, " Ariconium,"
on the edge of this district, was of like derivation—
Latinized.

THE TREE PIPIT.

Of late I have had excellent opportunities for observing
the habits of the Tree Pipit (*Anthus trivialis*)—a more
interesting bird than its sober plumage might bespeak it.
I think there can be no doubt of its being a connecting
link between the wagtails and the larks; its shape,
gait when on the ground, the nature of its food, with the
quaint vertical vibration of the tail, likening it to the
former, while the colour of its plumage and markings, but

above all the *timbre* of its voice, show its affinity equally
near to the latter. In some of its habits it is wonderfully
like the woodlark, especially that of perching on the
topmost twigs of a tree—usually a tall one—thence soaring upward while it sings. The song is neither so sweet
nor varied as that of wood or skylark, yet unmistakably
like them in tone ; so much so that one hearing it, without seeing the bird, would know he was listening to a
songster allied to the *Alaudinæ.* Nor is its flight either
so high or prolonged as theirs. It shoots rapidly upward,
in a line nearly direct, and at an angle of forty-five
degrees, but only for a distance of some sixty or seventy
yards. There, soaring for a few seconds, singing all the
while, it comes back to earth in a spiral curve, or, more
correct to say, to the top branches of a tree, though not
always the same from which it started off. In the last
twenty or thirty yards of its descent it exhibits a shape,
and gives utterance to a note, both peculiarly interesting.
The wings are at first widely extended, as is also the tail,
without beat or other observable motion, they are then
gradually drawn in towards the body, till the bird, seen
against a clear sky, in shape resembles the head of an
arrow, the wings representing the barbs ; and while thus
it utters a plaintive piping note, a very cry of distress,
some ten or a dozen times repeated.

Although more of a tree-percher than its near congener, the meadow pipit (*A. arvensis*), it seems to affect
places in the proximity of water, further likening it to
the wagtails. A pair have just brought forth young in a
tract of rather marshy pasture some two hundred yards
from my house, the nest being at the bottom of the grass
under a bunch of rushes. It was not found till the young
birds were nearly full fledged, then only three being in it.

On the finder returning to it some hours afterwards, there were but two, one of them badly hurt, apparently from having been trampled upon by one of the browsing cattle. I had them brought up to the house for examination, and while out upon the lawn inspecting them—an interval of nearly an hour having elapsed—I saw a little bird drop down upon the grass beside me, on the smooth, closely mown, and finely rolled sward, a bird of a species never observed there before, a titlark by its strut and the wag of its tail—a Tree Pipit—and it was the mother of the little fledglings I was in the act of examining.

There might seem nothing strange in this, but there was more than one thing strange. One, in the bird of a species which usually keeps far afield coming so close to the house and me. Another in its knowing where to find its young, abstracted from the nest. This was at least 150 yards off, with a thick grove intervening; and the boy who brought the young birds carried them under cover, so that they could not possibly have been seen by the parent. Nor could their tiny " cheep," uttered at intervals, have been heard by her; it was not audible to me at a rod's distance. How then came she to know of their changed whereabouts ? The only explanation I can think of is that seeing the lad take up what remained of her offspring, she had watched whither he went, and missing them from the nest, after a time repaired to the place in search of them. But that could not be *instinct ;* instead, something higher—surely an exercise of *reasoning !*

There is more to come concerning this little episode of bird-life, other incidents and observations yet incomplete, which, when completed, I may have an opportunity of laying before the reader.

THE NIGHTINGALE.

It is a common belief, even among British ornithologists, that this interesting bird does not find its way so far west as our western shires. In the latest edition of "Chamber's Encyclopædia," a work usually correct in points of natural history, it is stated that the Nightingale, " though plentiful in some parts of the south and east of England, does not extend to the western counties." Now, Hereford is surely a western county, and I can answer for it that at this present writing Nightingales may be heard every night, making Penyard Wood vocal with their matchless melody. Its western limit seems to lie somewhere near the longitude of Hereford city itself, and does not reach either Radnor or Brecon. For in lists of birds made out by competent observers, covering districts by the Welsh border, I see no mention of the Nightingale, and its presence in any part of Wales has not yet been chronicled. On the western side of Monmouthshire it is, I believe, also unknown, though where the Wye cuts through the carboniferous rocks in this county, in the valley of the river itself, the bird is a visitor. Independently of geographical range, it is capricious, or rather it might be called fastidious, in regard to the topography of its haunts and *habitat*. As, for instance, while Nightingales may be heard on one side of a hill, or range of hills, they will be silent on the other—in other words, they are not there. Penyard is an elongated ridge, full two miles in length; and often, returning home at a late hour of the night, around its southern slope I have heard as many as half a score of these birds in full song; no two together, but continued along the line of the ridge, each occupying a little ravine or section of its own, which

it seemed to have appropriated for the season. On the
northern side of this same Penyard I have never heard
the Nightingale, nor does it make its appearance there.
Moreover, it frequently affects one side of the river, while
shunning the other. I have friends living not five miles
off, but beyond the Wye, who will scarce give me credit
when I tell them that Nightingales sing all round my
house. They have never heard it on their side, and were
surprised to learn that the bird not only visits but breeds
in Herefordshire—their native county !

With regard to the geographical range of the Nightingale in our island, and the capriciousness above alluded
to, I have heard a theory advanced which seems worthy
of investigation. It is that the bird only frequents those
districts where the glow-worm is found. In the old red
sandstone of Herefordshire we have the *lampyris noctiluca*
in plenty ; and it is also abundant over the chalk formation of the Chiltern Hills, in Bucks, Berks, and Hertford-
shire, where Nightingales are most common. This seems
to favour the above theory, pointing to another fact—
that the luminous insect may be the favourite food of
the nocturnal songster.

THE MONTH OF BIRD-MUSIC.

May is usually accounted the month when birds sing
their loudest and sweetest. However this may have
been in times past, certainly for the last four or five
years June better deserves the credit. And never one
more than this now present. There were May-days preceding when wood and field, copse and hedgerow, were

GROUP OF WARBLERS.
177

all alike silent so far as concerned the singing of birds—not a stave of song heard in any of them; only occasional call-notes, or signals of alarm. How different all now, deep in the middle of June! This day (the 14th), driving out, and through wooded dells that border the Forest of Dean, though not in it, I had the pleasure of listening to a concert of bird-music, with so many voices taking part in it that to give the names of the singers would make a list large as ever sang on opera stage, choristers included.

Therefore, only one will I particularize—one well worthy of the distinction, the *prima donna*, not of the theatre, but of the grove. Had I ever doubted before that the nightingale sings by day, on this day my doubts would have been removed. At meridian hour, as before and after, with the sun shining brightly in a diaphanous sky, I heard its song, unmistakable as unmatched by anything else in the way of bird-music; and if there be any one sceptical of its singing by day, let him just now repair to the dells around the Forest of Dean, on the eastern or Gloucestershire side, and I promise him a change of faith.

AN OVERPRAISED BIRD.

" The male blackcap is inferior only to the nightingale in the quality of his song."

So asserts Mr. Yarrell, and the assertion has been repeated by all, or nearly all, ornithological writers since his time, till it is now generally received as axiomatic.

Yet never was statement much wider away from the truth. Not only is the blackcap's song inferior to that of the nightingale—with which it has no claim to comparison—but is beaten, far excelled, by those of thrush, blackbird, lark, linnet, and goldfinch.

I had often wondered at this concurrence of belief in the superiority of the blackcap's song, so different from my own impressions of it. But I think I have discovered the explanation. In nearly every instance where the naturalist of Selborne has made a mistake the error has been perpetuated by writers who have copied him; as, for example, that "crows go in pairs the whole year round." In the case of the blackcap's song, however, he has made no mistake; instead, described it with remarkable precision. His words are:—"The blackcap has a full, sweet, deep, loud, and wild pipe; yet that strain is of short continuance, and his motions are desultory; but when that bird sits calmly and engages in song in earnest, he pours forth very sweet, but inward melody, and expresses great variety of soft and gentle modulations, superior perhaps to those of any of our *warblers*, the nightingale excepted."

I have italicized the word warblers, for on that hinges the weight of White's opinion, which influenced his copyists, and so misled them. He clearly meant by it our summer visitants, the soft-billed birds, or *Sylviadœ*, specially known as the "warblers," without any reference to our permanently resident songsters. For of the former he was speaking when he so pronounced himself about the blackcap. Elsewhere he simply characterizes it as a "delicate songster," which is quite out of keeping with his entertaining the belief that of all our song birds it came next to the nightingale—as it certainly does not.

I hear the blackcap singing while I write, and through my window see the bird itself flitting about from tree to tree; for it is a restless creature, and does not remain long in one place, usually running over the whole of its gamut scale but twice or thrice, then betaking itself to some other perch, and there in similar manner repeating it. There are occasions, however, as the naturalist of Selborne quaintly expresses it, when "that bird sits calmly, and engages in song in earnest," its strain being undoubtedly one of the sweetest. No verbal or written description could come nearer giving an idea of it than that of White himself, his phrase "inward melody" having a peculiar and characteristic significance. While it is singing there is a muscular dilatation of throat, and erection of the crown feathers, forming a very distinguishable crest. Just now (last week of April) the blackcap is heard more frequently than later on, and oftener seen. When the trees are in full leaf, it is difficult to get sight of the bird, even when it is pouring forth its strain but a few feet from the spot where one may be standing.

NINE REDBREASTS IN ONE BROOD.

The singular ornithological fact which is elsewhere referred to is that of a pair of robins having brought forth nine young at the same hatching. The place of nesting was in the parish of Walford, near Ross, Herefordshire, and there was enough singularity in the time, the birds being out of the shell early in March. But nine of them, when the orthodox number is five, may seem something still more abnormal. There is an explanation, however, though even this leaves the occurrence one deserving to be called strange. There were in reality two nests but a few paces apart—one with five eggs, the other only four—and for a freak a school boy, who had discovered them, took out the four eggs and deposited them in the other nest with the five—there leaving them. As a rule, boys will not despoil the nest of the robin, and this urchin, being himself rather an odd and inquiring mind, made the transfer to see what would come of it.

What did come of it was that the owners of the five eggs continued incubation upon all nine, and in due time brought out the nine birds nearly together, fed and nurtured all without distinction, apparently unconscious of the trick that had been played them.

A WORD ABOUT THE SLOW-WORM.

As the Slow-worm (*Anguis fragilis*) is now also showing out of its winter quarters, it naturally attracts notice. Mr. Bell, in his " History of British Reptiles," the accredited standard work on our native herpetology, speaking of the Slow-worm, says that its " total length is from ten to twelve, or even fourteen inches." Why *even* fourteen inches ? Such loose, conjectural phraseology, too often indulged in by zoological writers, is likely to mislead, as in the present instance, when it gives an indefinite idea of the reptile's size—indeed, an erroneous one—which, after Mr. Bell, no doubt, has been copied and found a place in our standard encyclopædias. The error may be worth rectification, and I can rectify it from actual measurement of several Slow-worms taken in my own grounds, some of which were much above fourteen inches in length, and one actually exceeding seventeen inches. A curious habit of the creatures, half-lizard, half-snake, which they have in common with the true serpents, is their hybernating in bands of several individuals, twisted and coiled up together. A man in my employ, while working in a stone quarry, turned out such a cluster from a cleft in the rocks, where they had doubtless passed the whole of the winter. Awaking from their semi-dormant state, and separating, there were found to be sixteen of them.

A CURIOUS CASE OF BADGER-DRAWING.

In October, 1881, one of my friends out rabbiting sent two of his dogs into the burrow of a badger, hoping to draw the animal out. The hole, or " holt," as commonly called, was on the slope of Howle Hill, a lofty eminence overlooking the Wye, some four miles below the town of Ross. The dogs were fox-terriers, and valuable—one of them being a prize-winner—and, as time passed without either reappearing, my friend became anxious about them; all the more that no sound, neither bark nor yelp, came back out of the burrow.

Hours were spent waiting, with every effort made to coax the animals out. All in vain; neither call nor whistle received any response from the subterranean abode of the badger.

There seemed no alternative but to use the pick, spade, and shovel; which, in fine, were set to work with. As long trying-poles showed that the "holt" ran horizontally to a great distance, and laying it open from the mouth would be a task entailing great labour, it was determined to sink a vertical shaft instead. This was done by dint of hard digging, and the underground gallery reached, as it chanced, just midway between the two dogs, both of which were found dead. The one nearest the mouth of the burrow was jammed in a narrow passage, from which it had vainly struggled to extricate itself; while the other lay farther in, with open space around sufficient for turning, yet alike lifeless; but on neither was there mark of tooth or scratch of claw! The badger was also there, up at the extremity of the burrow, from which it was unearthed and killed.

Now, the question is, what killed the dogs? The one caught in the jam might have wriggled itself to death; but this hypothesis will not answer for the other, which had room enough to move about. And as there was sufficient atmosphere around to keep their lungs supplied, asphyxia will not explain it—unless it was produced by some powerful effluvia emanating from the badger. That this animal has the power of secreting a substance of most disagreeable odour, and projecting it at will, is well known; therefore the theory of the dogs being suffocated by it is not at all an absurd one—instead, plausible enough. If not, then how came they by their death? I can think of only one other cause—absolute fright at finding themselves hopelessly entombed. But that were still more improbable.

The badger was not one of the largest, scaling only 27 lbs. In my notes I have record of many weighing, at least, a third more.

A DOG AND FOX FIGHT ENDING
MYSTERIOUSLY.

Another instance of a dog entering a badger's burrow, and never coming out again, occurred just twelve months ago, and within a hundred yards of my own house—in the hanging woods of Penyard Hill, that rise directly to rear of it. The " holt " was, and still is, at the base of a cliff,—an outcrop of the old red sandstone conglomerate, —but in this case there was no badger in it; or, if so, he was not the object after which the dog was sent in. Instead, a fox had just entered, as told by its tracks in the snow; for it was during the long-lying snowfall of January, 1881. The dog was a rough Scotch terrier, that had often tackled both foxes and badgers ; and its owner, a poor man, supposing he had made a sure profitable find, urged it in after the fox, bagging the mouth of the hole, to secure the latter when it should attempt bolting out.

For some time the men outside—for there were two of them—heard the sounds of a struggle, a combat *à outrance* between dog and fox, as their angry voices indicated. But these gradually grew feebler ; not as if the strife were being relaxed, but carried farther away into the rocks ; at length ceasing altogether, or, at least, ceasing to be heard. Nor came there out any sound afterwards; neither issued forth dog nor fox; though for days the place was frequently revisited, and the snow carefully examined all around. Had either of the animals returned out again their tracks could not fail being seen ; besides, the terrier would have found its way home, the distance being only a few hundred yards.

In this case there was no thought of opening the burrow, which, being a natural cavity in the rocks, would have been a work of quarrying and cost. So the fate of fox and dog remains undetermined; though, certainly, it was death to the latter, and likely to both. But here again we have another mystery, difficult of elucidation as that which occurred on Howle Hill; the same question, under somewhat different conditions: what caused the death of the animals? Did they kill one another? Or did they go fighting on so far into the cavity as to be unable to find their way out again? Or was there a badger also within, that destroyed both as intruders upon its "holt" and home? Its outgoing tracks would not be seen, as it would not likely come forth so long as the snow lasted—too cunning for that.

A PROLIFIC POLECAT.

As is generally believed, the polecat, or fitchet (*Mustela putorius*), of which the ferret is erroneously supposed to be but a domesticated variety, is not so prolific as the ferret; yet there are instances of it also producing more numerously than is stated in zoological works. Mr. Bell, in his "History of British Quadrupeds," speaking of it, says: "The female polecat brings forth four, five, or six young." This may be the normal number; but I have note of a case in which it was exceeded, no less than *seven* young polecats having been dug out of a den, near the banks of the river Wye, all evidently of the same "kittening." This, in a way, tends to show near relationship between the ferret and polecat; and, beyond doubt, they are closely allied, yet still specifically distinct. As some proof of their being so, I may point to the close resemblance

between other species of *Mustelidæ* known to be distinct,
as between the stoat and common weasel.   These are so
graduated in size, the female stoat being little, if any,
larger than the male weasel, while so like in shape, *facial*
expression, and other respects, that were it not for the
stoat's bushy and black-tipped tail there would be some
difficulty in distinguishing the one from the other.   Mere
general resemblance in shape and colour is not enough to
justify specific sameness; besides, the polecat is usually
larger than the ferret, which would contradict the rule of
increase by domestication.   As for the white ferret with
pink eyes, it is a *lusus naturæ* of the "albino" kind, and
therefore not in the naturalist's category of species.

WILD FERRETS.

I think it highly probable that we have *Wild* Ferrets in
England ; that is, ferrets escaped from their owners and
living in a wild state—in short, become true *feræ naturæ.*
I am led to this conclusion by some cases that have come
under my own observation, with others reported to me.
It is well known that ferrets when sent in after rabbits
often remain inside the burrow, and have either to be dug
out, waited for, or abandoned.   When digging them out
is hopeless, from the nature of the ground, and to await
their coming forth inconvenient, they are frequently lost;
hence the cruel practice, still in vogue, I am sorry to say,
of stitching up their mouths, to prevent their indulging
in their bloodthirsty propensity when they have caught
the rabbit in the burrow.   The general belief is, that
these defecting ferrets are recovered again, either by being

found within a few days, or themselves returning to their owner, when his home is near at hand. Instances of the latter I am assured of; and also of a strayed ferret, whose owner had chanced to come across it in the woods, following him home as would a dog. But I am equally well assured of the other instances above referred to, where lost ferrets had not been found, and were still living months after having made their escape. As it is generally conceded that the tame ferret originally came to us from Africa, or the south of Europe, and is known to be " nesh " in cold weather, the supposition is that if left to itself in our woods and wilds, it would not survive the winter. But two of the cases that have come under my observation contradict this view, entirely refuting it. Some four years ago a man living in the parish of Hope Mansell, Herefordshire, lost a ferret while " rabbiting," and after trying his best to recover the creature, had to give it up. This was in early winter; and in the month of March following, when strolling through a track of woodland near the place where the ferret had got away from him, he espied an animal which he at first took for a fitchet (polecat); but getting nearer, by certain marks known to him, he saw it was his lost rat and rabbit-catcher. There were several other men along with him; and they immediately gave chase, running it from cover to cover, and hole to hole, routing it from each in succession, but still unable to lay hands on it, for it was as wild as any weasel. Nearly two hours were spent in skirmishing about after it; when, at length, one of the men, a labourer in my employ, who had stripped off his jacket, succeeded in throwing this over the animal, and so getting grip on it. It gave tongue, however—a harsh chatter—and teeth too, biting him severely. Now, this ferret had been out all the winter, and, moreover, an exceptionally severe winter; but that it had not suffered from the cold was evident, for, when recaptured, it was found in best form and condition, its coat sleek and glossy, itself fat, as if fed from an abundant larder.

Another curious circumstance I may mention relating to it. When caught, it gave out an offensive odour, of the true polecat essence, and quite as strong. Strange, too, that after being captured and restored to its hutch, it died within three days' time, though it had received no known injury while being "chevied" and taken.

Now, it seems only a fair inference that this ferret, having survived one winter out of doors, would have equally got through another, and another—in short, lived out the term of its natural life in the woods of Hope Mansell, had it been left to itself. And why not? As is well known, all animals of the weasel fraternity can go long fasting, if such be a necessity; though in the case of this ferret it seemed not to have been. And just for the same reason warmth would be within its reach, no matter how cold the winter, since it could lie up for long spells inside the burrow, and in the snug nest of a rabbit. Therefore, I conclude that there may be many ferrets living wild in our woods—" fitchet ferrets," as they are called, on account of their colour, and for this reason mistaken for fitchets themselves.

I am able, also, to record a case of the white ferret, which is still more intolerant of cold, running wild and outliving the winter. On the Warrage farm, lying contiguous to Raglan Castle, Monmouthshire, an old and large-sized "hob" of this variety escaped from its owner by getting into a long covered field drain. For twelve months after—and, therefore, the whole round of the year—it was seen at intervals in different places, and chased, but always managed to escape its pursuers. What became of it eventually is not known; but, no doubt, from its conspicuous colour, it fell a victim to the shot of some ten-shilling licensed gun. For it is not likely that the cold killed it, since it had already passed through the rigours of winter unscathed.

––––––

A DANGEROUS TRAP FOR TERRIERS.

Under the heading, "A Curious Case of Badger-Drawing," above, I gave an account of two fox terriers sent into a badger's burrow having to be dug after, and when reached, both found dead; the badger being close beside them alive, but, of course, killed by the diggers. In the next entry I further recounted another incident, where a Scotch terrier entering a badger's "holt" in a cliff behind my own house, after a fox which had taken shelter in it, neither dog nor fox ever coming out again. Instances of terriers being lost in this way are far from rare, and I have now another to chronicle, the particulars of which have been furnished me by one of my friends, an eminent M.F.H., who hunts one of the Welsh bordering shires. It occurred at the commencement of the last hunting season, and I give the account of it in his own words, quoted from a letter lately written to me :—

" The first day we were out with the hounds (I think 25th October), we ran a fox to ground after a long day's run. The terrier, Old Cæsar, as good a one as ever ran, got in after him; and though we waited and dug till dark, there were no signs of him or the fox. Next morning I sent over early, and they dug most of the day, and found Cæsar between a dead fox and two living ones! The two heads have been stuffed and mounted on one board."

This occurrence is all the more inexplicable from the dog and one of the foxes being found dead, while the two others were alive. Might it be that the terrier and chased fox, after a long exhausting run, were smothered in the hole by their own hard breathing? If not, how came they dead?

————

TOO TOUGH FOR EVEN A BADGER'S TOOTH.

In the same letter my fox-hunting friend gives account of another incident; curious, too, but this time more comical than serious. Thus runs it :—

"On the 12th of December we found, and had a good run, though ringing, and fox put into a drain close before hounds. The terriers soon bolted what I at once saw was a fresh fox; but hounds viewed him, and ran him to ground by the side of the river. I at once took them back to hunted fox, and found that two hounds and terriers had killed him in the drain. On opening it, a badger was also found in the drain; beside it a *china* egg, no doubt taken from a farm close by."

No doubt it was taken from the farm, and by the badger, for how otherwise could it get into the drain? The egg was one of those in common use as " nest eggs " ; and, as is well known, these animals prowl around farmsteads by night in search of *real* eggs, not counterfeits, and chicks as well. The curious part of it is the badger not discovering the counterfeit till he had carried it into the

drain.   Then he must have done so, finding it a "nut too
hard to crack," notwithstanding his sharp teeth and
powerful leverage of jaws.

BIRDS AND THEIR NURSLINGS.

Among the small birds there is a remarkable difference
in the mode of tending and feeding their young, which I
have just had an opportunity of observing.   I have elsewhere spoken of a tree pipit, whose brood was brought
to me for examination, the mother, in some mysterious
way, finding whither they had been taken, and, after a
time, appearing upon the scene.   I fancied the cock also
came, as a second bird, resembling a pipit, was observed
hovering about; but if so, he went off again, and was
not seen afterwards.   The hen, however, true to her
maternal instincts, stayed by her imprisoned offspring,
approaching as near to them as she thought safe, at
intervals uttering a tiny "cheep" of solicitude, to which
the youngsters gave response in much louder tone.

Placing them upon the grass, I withdrew to a distance
to note the result.   And a curious spectacle it was—the
manœuvring of the mother to get them away from what
she must have supposed a dangerous proximity.   Alighting on the ground, some distance beyond them, she
would run up till near enough for them to see her.   Then,
as they fluttered towards her,—for, being almost fledged,
they could do this,—she would turn tail on them, and
draw off a little way, again to make stop till they came
up.   This manœuvre was repeated time after time, till
she had coaxed them half-way across a field, in the direc-

tion of the nest. But, wishing to make further observations as to her mode of feeding them, I had the little fledglings brought back, and put into a cage, where they were kept through the night.

I have said there were but two, one having received some injury from being trampled on by a cow. In the morning this one was found dead; the other lively and active. The mother was flitting about in the neighbourhood, and had evidently fed it. The cage has a projecting shelf running all around its bottom outside; and as I watched her, she lit upon this with a large grub held crosswise in her beak. In a trice it was passed through the wires into the open mandibles of the youngster, when she flew away, and was for a time absent. Only about ten minutes, till she returned again, grub in beak as before, and, as before, gave it to the young bird—repeating and continuing the supply at intervals of ten or fifteen minutes throughout the whole of that day. And the same through several days after; for I kept the nursling some time encaged.

I was surprised at the quantity of grubs it managed to gulp down, in a single day devouring a bulk of them that must have been as big, or bigger, than its own body! And they were eaten alive, as many that "missed fire," from the difficulty of the mother getting them into its mouth through the wires, had fallen to the bottom of the cage, and were there crawling about without sign of damage done them. In the act of transference from beak to beak, I observed no attempt at killing or crushing them; indeed, the soft bill of the pipit would hardly serve for that.

The dropped ones gave me an opportunity of seeing that they were not all of the same sort, but of different species, differing also in size. There were some from the oak, others from the apple-tree, still others from the hawthorn; but the bright green caterpillar of the gooseberry-bush was more numerous than any; while a long-bodied black fly, of a species unknown to me, formed part of the varied diet designed for the all-devouring chick. No doubt it was having extra rations—all the provender that would have been supplied and otherwise shared by several deceased brothers and sisters, killed by the cow. I noticed that the flies, several of which lay at the bottom of the cage, were all dead—this, no doubt, done to hinder their escape while being passed into the beak of the young bird; but, as already said, the insects in the larval state were all living, as if there was no such fear about them.

In the end my observations were cut short by the young pipit escaping from the cage, through the "turnstile" of the seed box, that had been left loose on its hinge. It was evidently shown the way, and helped out, by its painstaking mother; and I never saw either again.

Not long, however, did the cage remain empty. In an Irish yew close by was a nest of greenfinches, with young also, well-nigh fledged; and, curious to note their way of tending their nurslings, I had them transferred within the wires. The finch being eminently a graminivorous, hard-billed bird, I wished to compare its mode of feeding the young with that of the soft-billed insect-eater.

In the very first scene there was a notable difference in the behaviour of the two sorts. Though the greenfinch may be called a *home* bird, usually nesting near the house, the pair operated upon showed far more shy than the pipit, whose haunts are afield. It was a long time before

they would come near the cage; and, when they at length
did so, it was not to alight upon it till after many comings
and goings, now flickering around it, then flying off again.
In time, however, they got over their shyness, afterwards
showing less timidity than had the pipit. But a more
remarkable difference was in *both* the parent birds coming
after their offspring, and both bringing them food—the
one as often and as much as the other. When either
drew near, the caged youngsters would commence flap-
ping their wings, giving utterance to a note not unlike
the chirrup of young sparrows when near leaving the nest.
Altogether different was that of the old birds as they
made approach, being soft and plaintive; for it was only
put forth when some one drew near the cage, and they
supposed there was danger.

But the most notable difference I observed between
these two birds of distinct *genera* was in the mode of
feeding their young. While the pipit, as already said,
brought the caterpillers in her beak, and transferred them
direct and living to that of the nestling, the finches
carried whatever food they had for theirs in the crop;
thence delivering it somewhat after the manner of pigeons.
In all their comings and goings, I could see nothing in
their bills, either bud or grub. Moreover, their intervals
of absence were more prolonged; as though from having
the means of carrying a greater quantity it had taken
more time to forage after and collect it. I noticed, however, that the pipit frequently brought back two caterpillars at a time, and once three, all of different species,
as I could tell by their diversity in size, as in colour.

HUNTING THE MARTEN WITH FOXHOUNDS.

It is painful to think that by the ruthless persecution of gamekeepers the marten, or "marten-cat," as often called, is fast approaching extermination in the British Isles. Both species, *Martes foina* and *M. abietum*, are now so rare that the capture of a specimen of either is an occurrence so infrequent as to find triumphant record in periodicals devoted to naturalist lore. Considering the paucity of our indigenous four-footed *fauna*, it seems a pity that such a handsome quadruped should become extinct, and all through wreckless misconception on the part of game-preservers. Perish the game, or a portion of it, say I, rather than that these beautiful and interesting animals should get totally extirpated, as ere long they are likely to be. But the remedy is still in our hands. Being, as all the *Mustelidæ*, of a highly prolific nature, a protective statute would soon restore them to numbers again, enough to make them, as they once were, a feature of interest in our sylvan scenery. And for the destruction of pine or beech marten, as of eagle, kite, osprey, or peregrine, the penalty should be a heavy one. Were such an act passed, and rigidly enforced, we should yet have the pleasure of oft witnessing the graceful and majestic flight of our grand *Falconidæ*, or, in a stroll through the woods, observing the pretty "martlet" playing, squirrel-like, among the trees. Although both the species of our martens (which some naturalists, without any valid reason, deem only varieties) sometimes frequent treeless situations among rocks, the tree is their real natural home and *habitat*, a hole in it nearly always their breeding place. The pine marten more especially confines itself to timber, and appears to be the better climber, though both are eminently scansorial. Indeed, to talk of their *climbing* is to use a very unfit phrase, since these weasels are as much at home upon the branches as the squirrel itself, and can not only run nimbly along them, but spring from one to the other, even from tree to tree; a fact I believe not generally known, at least I have never met mention of it in zoological works. It is just for this purpose nature has provided them with such a development of tail, long and bushy as that of the squirrel; not prehensile, but for balance and guidance, as the train to a paper kite, or the pole of the rope-dancer. Aided by this, and other anatomical peculiarities of structure, the marten not only passes safely from one tree to another, but, if needs must, can spring off from the highest, down to the earth, unharmed, as though it had made the perilous descent upon wings. As is well known, this remarkable, and indeed somewhat inexplicable, feat is common to most species of squirrels. In the American forests I have witnessed it hundreds of times; seen these creatures precipitate themselves from the tops of trees nearly a hundred feet high, drop lightly on the ground, and without a moment's pause shoot off like a "streak of lightning."

That our martens can do the same, or almost as much, I have reason to know, from many instances in proof; among others, one lately furnished me by a friend resident in a western shire, answering certain inquiries I had addressed to him. As his letter gives some curious details of *hunting the marten with hounds*, I will lay that portion of it before the reader, quoting his own words. Thus writes he :—

"I am sorry to say it is not in my power to give

you much information about 'marten-cats,' as we
have not seen or heard of one in this county for the
last eighteen or twenty years. Before that time they
were always to be found in particular localities, away
from keepers and preserves; and my uncle (who hunted
the L—— hounds for forty seasons) used to hunt marten-
cats very early in the season with the young hounds, and
a few old ones, to teach them to 'pack' well. The scent
of a marten-cat is so strong that it is hardly possible for
hounds to lose it, and my uncle used to say that it drew
them together and taught them to pack well, so that
when they began fox-hunting later on it almost saved the
expense of an extra whip. Foxes were so scarce in those
days that we could not afford to go cub-hunting in the
early part of the season, or we should have had many
"blank" days before the end. Of course, now that foxes
are more plentiful, young hounds can be entered to the
legitimate scent at the beginning. We used to find the
marten-cats in large coverts, and it was a common occurrence for one to give the hounds a run of three or four
hours in a thick cover, the animal every now and then
taking to a tree. From this it would be dislodged by
some one climbing up to it, when it would run along a
bough to the outside end, then drop into the cover, and
away again, although perhaps twenty couple of hounds
might be baying at it under the tree. I have seen one
'treed' at least a dozen times before it was killed."

I question the correctness of my friend's conjecture as
to the marten being extinct in the shire of which he
speaks. Indeed, I have evidence of its existence in that
county, though not in his neighbourhood. In my own, I
am happy to say, it is far from being extinct, many
recent cases of its capture having come to my knowledge.

Only six years ago a poacher of my acquaintance killed a beech, or, as sometimes called, "stone," marten within less than a mile from my house. He found it while "rabbiting," his ferrets having run it out of a hole in a hedge-bank, and far away from woods. No doubt it had made an excursion thither on the same business as the poacher himself.

But in many of the fastnesses around the Forest of Dean I know that martens, if not plentiful, are yet in goodly numbers. One of the Forest keepers tells me that, five or six years ago, he used to see many, and shoot many, too, in the High Meadow Woods—a tract of the forest which overhangs the river Wye; and there is the skin of one stuffed and mounted in the house of a farmer in that neighbourhood, which very recently fell to a gamekeeper's gun. Again, a gipsy of my cognizance, who tents in all parts of the Forest, tells me that he and his tribe often meet with "marten-cats," which he affirms to be far from uncommon in the woods near Blakeney and Lydney, where there is some rather heavy timber. He says they vary much in colour and markings—a remarkable fact, if fact it be. But he has promised to institute a search, and procure "samples" for me, if possible. So I await the result of this Bohemian's "cat-chasing" with a very vivid interest.

THE HEDGE-THREADER.

In early spring, the season of pairing and mating among our native birds, one of the most silent of them breaks out into song, to continue it at intervals, but still sparingly, through the summer months. I speak of the so-called hedge-sparrow, or hedge accentor (*Accentor modularis*), though both the above trivial appellations, as well as the scientific one, seem to me not only inappropriate, but somewhat absurd. Sparrow it is not in any sense, having no relationship with the true *Fringillidæ*, and the clumsy title, " accentor " is equally ill-bestowed upon it, as also " hedge warbler," another of its names. Still another, " dunnock," is too local, and of too obscure signification ; " shuffle-wing " being better, as denoting a characteristic habit of the bird. " Blue Isaac " is one of its designations in the Wye Valley, the name having reference to the bluish tint of its plumage, in connection with its quaint ways. As this bird is somewhat of a favourite with me, I will venture to suggest for it a cognomen which seems better than any of the above; viz., " Hedge-threader." No one who has ever watched it as it worms and threads about through stoles and branches in a hawthorn hedge will deny the appropriateness of the suggested title.

The song of the Hedge-threader—I decline calling it either sparrow or accentor—though not loud, is remarkably sweet; the bird, while giving utterance to it, standing perched on a spray, with open beak and shivering wings, seemingly straining upon its legs, as if the song cost it an effort. Not for its melody, however, does it so much deserve being a favourite as for its quiet, unobtrusive ways, and the confidence it shows in man. Even the robin itself is not tamer or more familiar round many a homestead. Still the Hedge-threader has its faults, slight imperfections of character ; for though a soft-billed, insectivorous bird, it is also graminivorous, and just now does considerable damage in the seed-beds of the kitchen garden. But for this it gives compensation, and far more, by destroying swarms of other seed and leaf consumers.

Viewed from a distance, the Hedge-threader appears a bird of sober, even sombre, hue. But take it in your hand, and you will discover a pretty mottling of colours, which, though dull, by their pleasing, regular arrangement, combined with the smooth trim set of its feathers, go far to redeem their want of brilliancy. This bird is one of those subject to erratic colouring of plumage, or, as commonly called, *Albinoism.* A specimen I am possessed of is of a beautiful buff from beak to tip of tail, with an edging of white on the wing primaries and secondaries, as also on the outside tail feathers. A very handsome bird it is, and no one not told would think of its being a " Blue Isaac."

Coming to the Hedge-threader's eggs, if splendour has been denied to the bird itself, these have it bestowed on them to an incomparable degree, as every nest-robbing boy but too well knows. There are few prettier sights in nature than the nest, with its precious treasures, rivalling the best blue of the turquoise.

CUCKOOS AND WAGTAILS.

The cuckoo appeared in the Wye Valley on the 11th of April. It may have been there earlier without my observing it; but on the afternoon of that day I saw a pair flying about not far from my house, at first taking them

CUCKOO.

for kestrels, as they gave out no note; but on nearer view I recognised the veritable *Cuculus canorus.* It is reported as having been seen, or, rather, heard, weeks earlier elsewhere; the truth of which report I am inclined to discredit, since the cuckoo's call is easy of imitation.

Much has been written about the variety of birds

which the cuckoo befools for her own purposes of procreation, and certainly the species are many, but all more
or less insectivorous.   Were it not so the young cuckoo
would have food given it on which it would poorly thrive,
or, rather, starve outright.   Around my neighbourhood
the bird it chiefly selects to do its hatching is the grey
wagtail, and yet the latter is by no means plentiful there,
save in certain limited localities; while we have the
cuckoo in remarkable abundance.   Some way or other
these find enough wagtails' nests to serve their ends,
though for a pair of cuckoos it needs more than one.   I
have note of four such nests around the same farmstead,
each with a cuckoo's egg in it, and certainly laid or
deposited there by the same bird.   Although hatched and
nurtured separately, and by different foster-mothers, I
think there can be no doubt about the cuckoo producing
several young at or about the same time, and that when
fledged and able to fly the individuals of this odd family,
nursed apart, become united under the guardianship and
tuition of their parents, remaining so till the hour of
autumn emigration.   Of this fact I had satisfactory
evidence in the after-summer of last year, by seeing six
cuckoos in a gang, four being young birds, as could be
told by their colour and markings, so different from the
old ones, the other two evidently their parents.   And
several days they kept together about my grounds,
unmistakably in family association.

A NURSE UPON THE BACK OF HER NURSLING.

Still another note anent the cuckoo and wagtail, furnished by my friend Colonel R., who is resident near me. Some years ago, stepping out upon his lawn, he was surprised to see a hawk, as he supposed it, with a wagtail sitting perched upon its shoulders. Drawing nearer, however, he discovered that the supposed hawk was a young cuckoo, and the wagtail, its foster-mother, feeding it. Watching them for a time, he saw the latter go and come, at each return bringing grub or worm in its beak, and transferring it to that of the voracious young monster, who ill deserved to be so assiduously catered for. On several occasions afterwards Colonel R. was witness to a repetition of this curious spectacle; and alike on the following year, the wagtail, as he supposed, being the same, the cuckoo, of course, different, but likely a younger brother or sister of that the beguiled bird had taken such pains to nurse on the preceding year.

THE ROOK AN OBSERVER OF THE SABBATH.

A clerical friend, a rector of long experience, who has given much attention to the habits of rooks, tells me that these birds quite understand the difference between Sundays and week-days. He speaks more particularly of those that breed about churches, and their behaviour, noted by him scores of times, is fair proof of the fact, however singular it may seem. Shy enough during the other days of the week, on Sundays they will be comparatively tame, permitting nearer approach, as though they knew that on the Lord's day there was no danger of their being molested. I myself have noticed their air of fearlessness, or trusting confidence, on this day greater than on others, and have no doubt of the fact. But how is it brought about? Sagacious bird as is the rook, its sagacity can hardly be equal to counting seven, or keeping a calendar. That it can tell a gun from an umbrella or walking-stick, or farm implement, is a fact well known; but its being able to distinguish Sundays from week-days is a still greater stretch of reasoning intelligence.

My friend offers an explanation, which is, no doubt, the true one: that the birds are made aware of the sanctity of the day, or rather its safety to themselves, by the ringing of the bells, and the assembling of the people for worship.

It would be worth noting whether they also lay aside their shyness on occasions when there is a funeral, or week-day service in the church.

WHY DO ROOKS BUILD BY CHURCHES?

In relationship with the fact of the rook distinguishing between Sundays and week-days is another of almost equal singularity—their choosing trees in proximity to the church as a nesting-place. For that they show this preference seems unquestionable. Proof of it may be seen at many country churches, where there are rookeries established on scant half a dozen trees of no great height, and easily accessible to the bird-nesting boy; while in the near neighbourhood are clumps of tall ones, just the sort one would expect rooks to build upon, showing not a nest. Nor can it be shelter that rules the selection. Often the trees by the church are in exposed situations, and the nests blown off to their last stick during the autumn equinoctials; whereas on other trees, only a few hundred yards distant, they would have remained throughout the winter with but little damage done, and so saved the labour of their rebuilding in spring.

It would seem, then, as though these birds have a knowledge that proximity to a church affords them protection, which it usually does, both from gun and nest-robbing boy, partly from the force of public opinion and, at times, fear of the vicar.

THE JACKDAW'S CONNECTION WITH THE CHURCH.

Whether the jackdaw be also a Sabbath observer I cannot say, but its connection with the church—the highest high—is unquestionable, and even closer than that of the rook. The attachment in its case, however, is of easy explanation, though it seems to puzzle the author of a book called "Wild Life in a Southern County," who makes it a text for much philosophizing, as follows:—"How came the jackdaw to make its nest on church towers in the first place? . . . Archæologists tell us that stone buildings of any elevation, whether for religious purposes or defence, were not erected till a comparatively late date in this island. Now, the low huts of primeval people would hardly attract the jackdaw.

It is the argument of those who believe in immutable and infallible instinct that the habits of birds, etc , are unchangeable, the bee building a cell to-day exactly as it built one centuries before our era. Have we not here, however, a modification of habit? The jackdaw could not have originally built in tall stone buildings. The jackdaw was clever enough, and had reason sufficient, to enable him to see how these high isolated positions suited his peculiar habits; and I am bold enough to think if the bee could be shown a better method of building her comb she would in time come to use it."

Does this writer not know that jackdaws breed also in cliffs, where they unquestionably bred before churches or high dwelling-houses were built? So, too, does the swift, swallow, starling, and some other birds which have also taken to church towers and other tall buildings as well. But what is there remarkable in this, or where the adaptation to changed circumstances? Some modification, it is true, but nothing more. These birds saw in the church tower, castle's keep, and chimneys of high houses just such places of security as the cliff afforded; hence their selecting them as a *habitat*, without any change of habit worth commenting upon.

As to the writer's analogy about the bee, though professedly conjectural, there have been facts recorded of this insect proving on its part a much more remarkable adaptation to changed conditions. Of all created things one would suppose it to move along lines limited by natural laws, with habits unalterable. Yet is it on record that a hive of Ligurian bees, carried across the Atlantic to tropical South America, and there set up as colonists, in the first year produced full honeycombs, in the second only half full, and the third none at all! The sagacious insects had discovered that in a land where "the flowers never fade, and the leaves never fall" there was food provided for them throughout the entire year, and no need of their toiling to lay up store of it.

A ROOKERY IN RUINS.

Returning to the rooks. These birds, however otherwise cunning, do not display this quality in the construction of their nests, which are so unskilfully put together as often to be partially or wholly blown down soon as built. Even an entire rookery has been known to go "by the board" under a spring equinoctial. Such a case occurred some years ago with a rookery belonging to a gentleman of my acquaintance in Worcestershire. The birds had nearly or altogether finished building, when a blast came that swept every nest out of the trees, scattering the sticks in litter all over the adjacent ground. The owner of the rookery was present to witness its ruin, and describes it as one of the oddest spectacles he ever beheld; from the forlorn, dejected air of the birds, as they sate upon the branches in clamorous council, some cawing loudly and in seeming anger, others in tone of doleful lamentation, just as human beings might act under a kindred misfortune. Indeed, their whole behaviour reminded him of the latter, the resemblance so quaintly comical that he, and others with him, could not keep from laughter.

A PLAGUE OF INSECTS.

The summer of 1880 was prolific in insects of the troublesome tribes. Short as was our apple and pear crop, it was still further minimized by swarms of wasps, these handsome but pestiferous creatures abounding everywhere. In an acre of orchard, with an undergrowth of clover and rye-grass, nearly a score of their nests were found at mowing time, the mowers and rakers having much difficulty in getting on with their work, and more than one of them coming in for a swollen head. A youth handling the hay-fork had one of his eyes "bunged up"; but, by the simple application of sweet oil to the part stung, the pain was allayed, and the swelling soon disappeared.

The nests were destroyed, with as many of the wasps as were "at home,"—not solely from spite, or vengeful feeling towards the insects, but more for an economic reason. The combs, or rather the cream-coloured *larvæ* contained in them, are regarded as the finest of poultry food, especially nourishing to young chickens; and, as most of the haymakers had hatches, this treasure-trove was eagerly appropriated and carried home to their cottages. My kestrels ate them with avidity; and it is from this the so-called honey buzzard (*Buteo apivorus*) has obtained its mistaken cognomen.

The mode of taking the wasps' combs usually practised in these parts is to insert a piece of quarryman's fuse into the cavity of the nest. The fuse, set on fire, is covered up with a sod, or shovelful of earth pressed hard, to prevent the issue of the sulphurous smoke, which pervading the cavity, destroys the insects. The time usually chosen for the operation is after sundown or late twilight, when the wasps have returned from their wanderings and gone to rest for the night, because then the job can be done with less danger of being stung by them. The comb-gatherers universally assert that a single individual of the hive, much larger than the common kind, and which they call " the main wasp," is always found keeping sentry at the entrance of their subterranean dwelling. Despite severe frost, these yellow gentry are still active among my pears, though not in such numbers as at an earlier period.

I do not remember an autumn in which the harvest bug (*Leptus autumnalis*) has made itself more felt. Seen it is not, or very rarely, since only sharp eyes, actually searching for it, may detect its presence. When seen it reminds one of a minute particle of Cayenne pepper more than anything I can think of, for it is just the colour of *Chilé colorado*. It is exceedingly like the *chica* of the Mexican tropic-land, better known as the "jigger,"—a corrupted synonym of the West Indian negroes,—and it is certainly the British representative of this dreaded little beast. The jigger, however, usually confines its attentions to the feet and toes, while the harvest bug ranges higher, inserting its poisonous proboscis into the ankles and legs, up to the hips. The inflammation produced by its bite, or sting, if not painful as that of the wasp, is far more prolonged, lasting for days, and, alas! also nights, the victim of it often tossing to and fro for hours on a sleepless bed. The torment is over now, with the season for its activity; but many a skin will still show purple spots—souvenirs of its baneful presence. Specially affecting wooded districts, it is nowhere more plenteous than on Wyeside.

Just now another insect pest has replaced it, also of

P

minute dimensions—so small as to be barely visible. A winged insect this—the little grey midge, which here and there hovers in swarms, as gnats. Unlike the harvest bug, it directs its attacks against the upper extremity of the person, alighting upon, and biting, the exposed parts of the neck and face, more especially the ears and forehead. It will even penetrate through a close crop of hair, and make itself felt on the skin of the head, if hatless. The irritation is such that the finger-nails are freely used, till "bumps" show all over the head and behind the ears, the ears themselves red from repeated rubbings. Even the weather-hardened cuticle of the rustic is not proof against its envenomed darts; and labourers engaged in outdoor work, when under trees, or in the farmyard, are often grievously annoyed by it. This year the midges are more than usually numerous, and more than ever virulent.

So also have been two species of large flies that afflict horses and cattle. "Have been," for luckily the cooler atmosphere has long since routed them. One is the forest, or horse, fly (*Hippobosca equina*), an ugly brown insect; the other of bluish colour, locally known as the "bree fly." Cows bitten by the bree will throw up their tails, and gallop about, bellowing as if mad, and breaking through fences they would not otherwise face; while one fastening upon a horse, especially if a fine-blooded, thin-skinned animal, will cause him to bolt in harness, if the reins be not dexterously handled.

THE WOOD MOUSE.

The Wood, or long-tailed field, Mouse (*Mus sylvaticus*) is quite common in this neighbourhood. The naturalist of Selborne seems to have been unacquainted with it; therefore I suppose it must either be rare in that district of country or altogether absent from it. Here it is too plentiful, having this year done some damage to my potato crop, in the digging of which, the other day, my men turned up a nest from among the weeds and haulm. It was empty, the young, full-grown, having gone out of it. But remaining in proximity, two unfortunates fell

LONG-TAILED FIELD MOUSE.

victims to the ruthless diggers, who never allow animals of the order *Muridæ* to escape. On examining the lifeless pair, I found one to be an "old buck," no doubt the father of the family; the other a young individual, of the same sex, with like certainty the son. The old mouse measured seven and a half inches from snout to tip of tail—the tail being exactly one-half, or just the length of the head and body. The squirrel or dormouse colour, which the Selborne naturalist speaks of as characterizing

the harvest mouse (*M. minimus*), is also a characteristic of
this species over the upper and back portions of its body,
as is the fine silvery white below, and the well-defined
horizontal line separating the shades.   Indeed, White's
description of the harvest mouse will stand good for the
Wood in all save dimensions, both differing essentially
from the third British species—he of the house (*M.
domesticus*).   From it they also differ much in habits,
while in these closely assimilated to one another.   The
nests of both are of spherical shape—a hollow ball of
grass without visible entrance; that of the lesser species
being of smaller size, but firmer.   The leaves, not the
stalks, of grasses are employed in this curious construction, those on the outside being broader and coarser, the
lining composed of blades which have seemingly been
split to render them finer, as there is evidence of gnawing
and tearing at the ends.   As is well known, the nest of
the harvest mouse is placed high up on the stalks of
thistles, wheat, or other corn, attached to and supported
by the culms; that of *M. sylvaticus* resting by their bases,
though still above the surface of the ground.   The two
species are not alike prolific, the smaller one producing
as many as eight to the litter, the larger never more than
six; at least, in several nests examined by me there were
but this number of young.

The legs of the long-tailed mouse, as seen in the specimen before me, are almost snow-white, and the length of
the hind foot nearly an inch.   This points out a peculiarity, an affinity with the squirrels and jerboas—the
power to sit erect on its hind quarters—which it has.   It
is altogether a handsome quadruped, larger than its
domestic congener; while its ruddy colour, and grand
black, glistening eyes, with ample high-peaked ears, give

it an aspect very different from that so repulsive in some other members of the family.

I have seen it somewhere stated that the Wood Mouse, as also the harvest species, occasionally eats insects. I doubt there being any truth in the statement. Its dentition is essentially of the rodent character, and my "Mexican" potatoes, gnawed here and there around the neighbourhood of the nest, give proof that the pair taken have been for some time past indulging in a diet purely vegetable. In their stomachs I could detect nothing recognisable save by chemical analysis.

TREES LEAFING TWICE IN THE YEAR.

Unobservant people may think it strange when told that many, indeed, most species, of our deciduous trees in certain years produce what might be called two crops of leaves. Not of themselves, naturally, but forced to it ; though the forcing is also due to nature, through the *larvæ* of insects feeding upon, and often totally destroying, the first output of foliage. The spring of 1881 in many places gave striking illustration of this, whole patches of woodland, especially oaks, throughout the month of May showing leafless as in midwinter. But the "midsummer sap"—for it is this which renews the foliage—brought them about ; and in after summer they were again green, with a leafage as full and luxuriant as those which the caterpillars had left unscathed.

TREES OPENING THEIR LEAVES AT DIFFERENT DATES.

With trees the time of budding and leaf-expanding, of course, depends very much on the species; some, as the chestnut and willow, being much earlier than others, as the beech and oak. This every one is aware of; but it is not so generally known that trees of the same species put forth their leaves at irregular periods, with days, sometimes weeks, between, even when growing side by side in the same copse or wood. Just now I have an instance of this before my eyes, in a hanging wood which forms a background to my house. The trees in it are for the most part oaks; and, strange to tell, some of these are now (April 17th) nearly in full leaf, while others show bare branches, or only with inconspicuous buds on them. I am inclined to think that these earlier leafers are the ones which last year suffered devastation by the caterpillars, and had to put on a second dress. Judging by their place in the wood it would seem so; but, unfortunately, I made no exact note of this, and about such a matter memory is not to be relied upon.

THE FLOW OF SAP IN TREES.

Glancing into a novel I chanced lately to lay hands on —"Vixen," by Miss Braddon—I was rather amused at reading as follows:—

"The moon had risen, a late October moon. . . . Here and there a sturdy young oak, that had lately been stripped of its bark, lay among the fern like the naked corpse of a giant. Here and there a tree had been cut down, and slung across the track ready for barking."

Reading Miss Braddon's books, one is disposed to believe her almost omniscient; but if this be a specimen of her knowledge, I fear it is not always reliable. Bark-stripping in October would not only be an anachronism, but a difficult operation; and, had Miss Braddon the "strippers" to pay, she would find it a costly one. But to the romantic writer, I suppose, there is nothing impossible. Dropping criticism, which is meant in no hostile spirit, I come to speak of bark-stripping, a business now, in mid-April, about to commence. Nor do I here intend giving account of the operation itself; only in its relation to one of the phenomena of nature. As all know, at this time of the year the sap in trees is fluent, or, as commonly expressed, "running"; which gives the bark-stripper his opportunity; otherwise the task of removing the rind would be well-nigh impossible. But perhaps few are aware of the fact that there are *three* runnings of the sap, or three "saps, as the strippers term them; their respective times of flow being quite distinct from one another. The earliest, or "spring sap," as called, is longest of continuance, lasting for a month or more, and is the one made most of by the strippers. A second flow succeeds later on, after an interval of stagnation; which is the poorest and deemed of least account for their purpose. Still later, about the last week in June, comes the "midsummer sap," of somewhat longer duration; when again the oak can be conveniently divested of its rough coat, and the stripper returns to his task for a short and final spell. But in favourable years his work is nearly, if not altogether, continuous, the three "saps" succeeding one another by intervals of only a few days,

THE ANNUAL FORAY OF THE PHOTOPHAGI.

Last year trees of nearly every sort suffered much from what is commonly called "blight," whole tracts of woodland, especially where oaks abounded, showing bare branches in May and June, when they should have had on their brightest livery of green. It is a popular belief that this *blight* is due to atmospheric influences—"something in the air," as I have heard country people say of it; and, in truth, the air has something to do with it, but not in the sense understood or fancied by them. The entomologist, of course, knows the real cause, in which there is no atmospheric mystery, but a simple operation of nature, though irregular in its workings, or, rather, the amount of work done by it in the different years. This irregularity alone is to be accredited to the atmosphere, the blight itself proceeding from the *larvæ* of certain species of insects belonging to the order of leaf-eaters, and chiefly of the family *Cynipidæ*. It is a numerous family, the oak itself being the foster-mother, as it were, to many of its members, the more notable ones being nursed in what are indifferently called "oak-apples" and "oak-nuts," but more properly "oak-galls." And I may here remark that the famed "Dead Sea apples" are similar excrescences, created by an insect of the genus *Cynips* on the leaves of a species of Syrian oak.

There are few insects whose life and ways are more interesting than those of the *Cynipidæ*, even the ants not excepted; and I hope, later on, to have an opportunity of giving further and fuller details about them, the present note being only meant to chronicle some facts which have just come under my observation. In the Forest of Dean

enclosures, not far from my home, I had heard that
there were places where "grubs" were hanging so
thickly from the trees no one could pass underneath without having coat and hat covered with them, the hideous
creatures also coming slap against the cheeks, and there
adhering, to the annoyance and disgust of the wayfarer.
On paying a visit to the place, I found things as represented, and that the suspended grubs were of *Cynipidæ*
in their larval form. On some of the silk-like filaments
on which they dangled, thin as a spider's thread, I
counted as many as a dozen, showing the great strength
and tenacity of this curious material. But they were not
all swinging about; instead, a number, and the greater
one, had descended to the earth, and were all over the
grass, evidently browsing upon it. Some young birches
that grew under the oaks were also thickly beset by
them, and I saw they were feeding on the leaves of these
as well—a proof that, as with *termites* and locusts, no
vegetable substance comes amiss to them. Several of
the young birches were already defoliated, others only
half stripped of their leaves, with the work of devastation
going on, and still others where it was just commencing.
Breaking off a spray from one of the last, and closely
scrutinizing it, I was able to make out no less than
eleven distinct species of these insect *larvæ*, and of nearly
as many different sizes—from that of a cheese-mite to
grubs over an inch in length. They were alike varied
in colours, too, green of several shades predominating;
though among them were none of the vivid green species
which affects the gooseberry bush. There were some
quite black, and others of a dull, dirty brown, all ugly
enough. And to watch them moving about over the
leaves and branches, in their peculiar jerking way, now

one standing upright on the edge of a leaf, or stretched out horizontally to its fullest extent, feeling about for other support; now two or three, and of different kinds, meeting, wriggling together, and crawling over one another—all this was an interesting spectacle, though not a very pleasant one. For no form of animated nature would be much more repulsive than that of the caterpillar. I could not see that there was any antagonism or hostility between the different species; indeed, all seemed on an amicable footing, and engaged in the common purpose of leaf-eating, to prepare themselves for the next stage of their curious existence—that of *chrysalis.*

A somewhat interesting fact in relation to these insect *larvæ* has been communicated to me by a man who keeps sheep in the forest. He says that in places where the grubs get upon the ground the sheep shy away from them, and will not touch the grass so infested; all of which is quite natural and comprehensible. And this leads to consideration of another fact, more difficult to comprehend, if, indeed, possible—that in a wood where these caterpillars appear, instead of scattering all over it, they do their work of leaf-eating in a regular way, taking the trees in belts, often with well-defined edges, just as do human beings at bark-stripping.

A CASE OF BIRD EVICTION.

It is stated by some ornithological writers that the starling occasionally takes possession of the green woodpecker's nest, evicting the owner by force. If the statement be true, then is the fact a strange one, since neither in size nor strength is the starling a match for *Picus viridis*, whose sharp pick-axe of a beak should be armour sufficient for either attack or defence against a far more powerful adversary. But I doubt the fact of this alleged dispossession, notwithstanding that an instance of starlings having appropriated the nesting-place of green woodpeckers came under my own observation. It was the same I have spoken of as in my orchard,* where the woodpeckers brought forth the brood that disappeared so mysteriously. This was in the summer of 1879 ; and revisiting it late in the following spring, to ascertain whether these birds had come back there to breed, I found the tree cavity occupied by a pair of starlings, who had nested in it, and were in the act of incubation. Left in undisturbed possession, they brought out their young, successfully rearing them, and again another brood in the succeeding summer—1881.

I might have believed it a case of forcible dispossession but for a fact which goes far towards contradicting this view of it, if not altogether disproving it. In the long-lying snow of January, 1880, a green woodpecker was found dead in the orchard near where the pair had nested, in all likelihood one of the old birds. If so, this would account for their non-return to the nesting-place, without the starlings having anything to do with it. Besides, it might be that, after all, my haymaking lad robbed them of their young, which would be sufficient reason for their never more caring to make nest in that apple tree, " under the mistletoe bough." So the starlings are doubtless innocent of having *evicted* them, and but took possession of a home they found untenanted and ownerless.

* Pages 43, 44,

A PAIR OF UNFORTUNATE BIRD MOTHERS.

Although among the family of *titmice*, or tits, there is much similarity in general habits, there are, nevertheless, some remarkable points of difference in what may be termed their *moral* characteristics. In most of their ways no two approach nearer one another than the great tit, or oxeye, and the little blue; yet between them I have of late witnessed an incident illustrative of these traits of distinction. Sitting out in my grounds some days ago, I observed a great tit fly into a hole in an old laburnum tree, which has got decayed at the heart. Approaching the place, as anticipated, I found there was a nest, and the bird sitting upon eggs. To ascertain their number, and whether she was in the act of hatching, or only laying another, I inserted the end of a rod into the cavity; when, after a little persuasion, she flew off, escaping by a lateral orifice in the bark.

The eggs proved to be nine in number; and after counting them I returned to my chair, and sat watching for the bird to go back to her nest, for I had ascertained that the eggs were all laid, and incubation had commenced. Instead of returning, however, immediately, as I expected, she remained absent; neither could I see nor hear anything of her. At the time there were some men doing garden work near by, who, seeing me so interested about the tit's nest, said they believed there was another in the wooden casing of a rain-water pipe, which they pointed to. This was in an angle of the house walls, about twenty feet from the laburnum, the top of the casing being nine or ten above the ground. It had a wooden cap, where a small aperture was observable, into which the men had seen the tit enter. A ladder

being brought and the lid lifted off, just under it a tit
was discovered upon her nest; not the *Parus major*, but
the little "nun" (*P. cærulous*). She was within six
inches of the boy's eyes who went up the ladder, and
had to be touched several times before she would move
off. This, however, she at length did, when the eggs were
counted—eleven. But now the behaviour of the bird
claimed my attention, so different from that of the congeneric species. Instead of flying afar off, and altogether
disappearing, she remained in the immediate neighbourhood, showing excited and solicitous about her egg
treasures, and proclaiming it by an almost continuous
utterance of her *cherring* note. The male bird was there
too, having joined her on the instant; and the pair went
flitting about from place to place, but still keeping near
the nest. As the wooden cap had been replaced, the
ladder removed, and every one had gone back to their
work, I looked to see the hen tit now return to her nest.
Which she did, but not till after many approaches and
returnings, in all occupying twenty minutes' time. But
still the other incubator had not come back to her nest
in the laburnum, nor could I see anything either of her
or her mate, though I remained watching for nearly an
hour; then left the place, having been called away from
it. Curious to know whether she was still absent from
her nest, I returned to it shortly after, to find her there
sure enough, close squatted over the eggs. This time
she was left undisturbed, and I had the satisfaction of
having discovered a *moral* difference between the two
species, evinced by the behaviour just observed.

Three or four days after, passing the laburnum, which
stands by the edge of a gravelled walk, I glanced into
the cavity, expecting to see the tit on her nest, this

being but five feet above the ground. Instead, I saw only the eggs, and supposed she was off them for a moment in search of food. But going back again some hours after, I noted that she was still absent; and, as my visits were several times repeated, with the same result, I came to the conclusion the bird had abandoned her nest. The eggs were there, all nine of them, but cold to the touch, as though they had not been lately sat upon. This, in fine, proved to be the fact; and now, knowing the nest abandoned, I broke one of the eggs, to ascertain how far they had been hatched. The embryo bird was in process of taking shape, which betokened an incubation of some days. But why had the mother forsaken her brood so soon to be? She had only been once disturbed, though several times looked at by passers-by. Was this the cause of her defection? For some time I supposed it might be, knowing that several species of birds have the habit, not only of deserting their eggs, but young, when the nest has been too often visited. As it turned out, however, the explanation seems to be different, my gardener three days after having found a dead tit on one of the walks, the hen bird of *Parus major*, no doubt the mother of the unhatched brood in the laburnum. But there is still a mystery unsolved,—as to how she came by her death,—since there was no wound nor other sign of injury—not a scratch of skin or ruffle of feather upon her! My narrative of these two incubatory birds, I am sorry to say, is not yet at an end, having to record a still more painfully tragical fate for the little nun. Wishing to ascertain whether the eleven eggs had been all fruitful, I had the ladder re-erected, and the boy sent up again. On lifting off the wooden cap, he saw the mother bird, as before, sitting on the nest, but in a somewhat unnatural attitude, a little away. At the touch she refused to fly off or stir; and, no wonder, as she was dead!—cold, stark, and stiff, with the eggs still un-hatched under her.

Now came the question, What had killed her too? Examining the body, I could find no wound, though there were traces of scouring around the vent. But what could have caused this? And, if so conditioned, why had she remained on the nest, seated upon her eggs—to die? The only explanation I can think of is that my servant, on the first occasion, replacing the cap of the wooden casing, had pressed it down closer than it was before, so narrowing the passage to the nest; and the bird, having squeezed herself in, was never able to get out again. I had noticed that she seemed to have some difficulty in effecting an entrance. The poor thing must have been dead for many days, no doubt dying by inches; a sad fate to reflect upon. But there is something even sadder to come. My gardener had told me that he several times saw the cock bird clinging to the head of the wooden casing, by the entrance to the nest, and tapping upon it with his beak; as the man supposed, bringing food to the hen inside, and so signalling to let her know it was there. The fact had greatly interested me; but, alas! I now knew that the tapping must have a different and more painful interpretation—the male bird knowing its mate, that should soon have become a mother, imprisoned, hopelessly shut up, as it were, in a living tomb!

A LAMB WITH TWO MOTHERS.

A somewhat curious case of sheep maternity, display-
ing instincts of the cross purposes kind, occurred in my
flock during the past year. About the middle of March, a
young ewe of the Welsh mountain breed—a yearling, and
quite black—gave birth to a ram lamb of the same colour.
As the yeaning came off in a corner of the pasture field
shaded with trees, where the ground was damp and cold,
I directed my shepherd to remove mother and young to
a drier and sunnier spot, about 100 yards distant;
which he did, taking up the lamb a few minutes after it
was dropped, and by a series of manœuvres coaxing the
dam to follow. She followed, showing great reluctance,
however; and after reaching the new ground, turned and
ran back to the place of parturition. This she did repeatedly, though coaxed away from it again and again,
till at length the lamb had to be left there with her.
And then occurred the first scene in a chain of incidents,
which I think may be pronounced not a little singular.
When the lamb, directed by instinct, approached to
suckle her, the mother would not allow it; and, instead
of showing the usual solicitude, absolutely repelled it,
butting it off whenever it attempted to take hold of the
teat. This she did over and over again, and her hostile
temper continuing, it became necessary to have her
caught and held while the lamb suckled her. To save
repeated chasings and catchings, I had her brought into
the ornamental grounds by the house, and there tethered;
the young one with her, but left loose. On its part there
was no lack of filial fondness, though still the unnatural
parent refused to give the nourishment due to it, and had
to be held every time it suckled her. And held hard,

too, as on each occasion she made violent struggles to escape.

Three days were passed in this forcing process, when, by chance, another yeaning ewe of the same flock and breed, but a white one, dropped a dead lamb, the lamb being also white. So, partly to prevent the swelling of her udder, as partly for experiment's sake, I had this white mother also brought upon the lawn, and tethered just outside the rope radius of the black one. Then the lamb was put to her, and although so different in colour from her own dead one, which had been with her some time, she not only suckled the blackamoor willingly, but appeared greatly pleased with it.

For several weeks the two ewes were thus kept picketted on the lawn at a little distance apart, the lamb running loose between them. And during all this time its black and real mother would not let it have a drop of her milk without being held, instead always "bunted" it off angrily; while the white foster-mother fondled and freely gave it all she had. Still the filial instinct remained true to nature, though the maternal one was false; and the little creature, despite all repulses, kept closer to, and seemed fonder of, its own unkind mother than the one that had so kindly adopted it.

Concurrent with this call on its divided affections, there were other claimants to a share in them. Being a beautiful creature, it was often taken up in the arms of a fair lady, and brought inside the house, where it made the acquaintance of a white bull-terrier, and a Persian cat of the same colour. In common with these it was allowed the run of both dining and drawing-room; and scores of times have I seen the three quadrupeds, types of an internecine hostility—tiger, wolf, and sheep

Q

—lying peacefully asleep side by side on the hearth-rug, with legs across and heads pillowed on one another. And as often the three playing at romps together, to the serious detriment of carpets.

There came a time when this must necessarily cease, by the lamb, alas! threatening to become a sheep; then it had to be relegated to its proper sphere, the pasture field. And, no longer needing sole sustenance from the teat, its two mothers were released from their tethers, and set adrift in the field with the rest of the flock. And now another odd incident of the series. Out afield and free, the foster-mother continued her affectionate attentions, standing for the lamb to suckle her, and caressing it the while. But the real mother, as all along, still repulsed it whenever it attempted to take hold of her teats; yet strangest thing of all, she would keep close to and run after the little creature, even to following it through the mazes of the flock! Watching their movements, day after day, I could not avoid the conviction that there was bitter jealousy between the mother and the nurse, though, so far as I saw, no fighting took place.

For awhile the petted lamb permitted itself to be caught; and when carried into the house would acknowledge its old canine and feline acquaintances, though no longer disposed to play with them. Soon, however, it became shy, indeed, wild as any of the other lambs, its new associates, and was caught up no more.

In due time it was made a wether, and as the rutting season approached, early in October, I had my white sheep separated from the black ones, and put into fields far apart. This, of course, parted the lamb from its nursing mother, leaving it with the real one, who still refusing it milk, it had to take wholly to the grass. The separation lasted for two months, till the last week in December, when the black sheep were brought back into the field where the white ones had been left. And now was I witness to another strange episode, the last of the series. Soon as the flock of black sheep entered at the gate, all "baaing" and bleating, as were the white ones inside, the two-mothered lamb—now grown sheep size—made a rush open-mouthed for the nurse from which it had been so long separated, and seized hold of her milk-less teats. But she now repelled it too, though not un-kindly, evincing by looks and gestures that she not only recognised her foster-son, but perfectly understood the situation !

A SINGULAR INSTANCE OF CANINE SAGACITY.

I have a sheep-dog whose sagacity is truly surprising; he seems up to everything short of articulate speech. But *think* he can, and clearly, as testified by the expression of his eyes, and the display of cunning with great capa-bility in his actions.

He is of a strain somewhat remarkable, the bitch, his great granddam, having borne whelps to a *dog fox*, one of which was his grandsire. This singular cross occurred among the mountains of Breconshire, in a wood adjoining the sheep-farm where the bitch belonged. And the bring-ing forth was in the fox's burrow, inside which the pups were suckled by their dam, and there kept till able to run about. Then these half-bred canines were caught and brought home to the farmhouse, the mother following. It was a curious instance of cross-breeding between the tame and the wild ; animals, too, so specifically distinct, besides, usually at bitter war with one another ! Still not unprecedented, many similar cases having been recorded.

Whether his semi-vulpine ancestry has done anything to sharpen the wits of my sheep-dog, I know not ; though like enough it has. Still there is nothing vulpine in his nature, no fierce or ravening instincts, as might be expected from such a strain ; instead, he is remarkably gentle and affectionate. And never so happy as when he sees a flock of sheep in the far-off field, and stands awaiting the order to fetch them to the fold or up to the foot of the shepherd. Then his eyes fairly dance in delight, his whole body quivering with anticipated pleasure. On getting the word " go," he is off like an arrow from the bow, or a greyhound unleashed at a hare. But not with like evil intent, for he treats the ovines tenderly as may be. Necessarily, now and then, with his snout, he bowls over one that is obstinate and will not run the right way, but never to bite nor tear it.

He is up to all sorts of sheep-dog doings, and that is being up to a great deal, since some of these are positively astounding. One I was witness to the other day furnished as clear evidence of mental ratiocination as could well be. A flock of sheep was being driven along the road with, besides the driver, two dogs attendant. One of these kept behind the sheep, the other in advance of them, and at each open gate or break in the bordering fences, the latter would take stand, and stay there as a sentinel on post of guard till the headmost of the flock were fairly up, with the certainty of their passing on. Then would the knowing animal start off, and rush ahead again, to look out for any other opening there might be

along the double line of fencing. Nor was this all; a still
greater degree of sagacity on the dog's part remaining
to be recorded—a very subtleness of reasoning, for to
call it instinct were to palter with words. When there
was a hole or "glat" in the fence, doubtfully big enough
to give passage to the body of a sheep, I saw the dog
stand regarding it, evidently pondering on the possi-
bilities of the sheep getting through, and at length,
satisfied they could not, trot on to examine the
next!

But Bob—as my own beautiful canine is called—can
do all this, with the other sheep tricks, and something
more; a thing I should have been loth to believe without
actually witnessing its accomplishment. As all know,
the tick is a troublesome pest to the poor sheep, oft
irritating them exceedingly, and a good shepherd will
now and then do his endeavour to rid them of the annoyance by picking the insects off. Several times when
mine has been so employed have I seen Bob helping
him; the dog burying his snout in the sheep's wool, and
nosing about till he came upon a tick; then catching and
"scrunching" it between his teeth, in a most business-
like manner, when he would drop the ugly beast, and
proceed in search of another!

I make no comment on this curious proceeding, further
than to say that, when I first witnessed it, I was struck
with astonishment. Who could have been otherwise?
And after that, who is the sceptic to deny to dumb
animals the possession of *intellect,* altogether apart from
instinct?

A BIRD'S INSTINCT, OR SAGACITY, WHICH NEEDS EXPLAINING.

In *The Live Stock Journal* I once gave an account of a tree pipit (*Anthus arboreus*) that had discovered its young in a cage where they had been put, after being carried off from the nest, and so transported that the parent birds could not possibly have seen whither they were taken. I can now record a still more singular case of a similar kind, the despoiled nest being that of

CUCKOO AND WAGTAIL.

a wagtail, and the abstracted bird a young half-grown cuckoo. The latter, taken away from the nest, where it was a usurper, was carried inside a farmhouse, into one of the rooms, there deposited in an empty blackbird's cage, and was for a time left to itself. He who so placed it, returning in an hour or two afterwards, found it no longer alone; instead with one of the wagtails, its foster-mother, clinging against the side of the cage, and feeding it! The window was open, and through this she had entered; but how it knew of the young cuckoo being there is the puzzle, for the place of the nest was some way off, and when the chick was taken out no wagtail appeared to be about. The only plausible explanation is, that some signal note may have been sounded through the open window, heard and mutually understood by the wagtail outside and the young cuckoo within. I may add that the wagtail came regularly afterwards into the room, and fed her foster-child till the latter was full grown.

A HERON TICKLED WITH A TROUT ROD.

One of my friends, an ardent disciple of "Izaak," with home near Abergavenny, gives me relation of a curious incident that occurred to him some time ago. He was out angling in the Usk, and while working along the river's side came upon a small but deep inflowing stream, a brook with high banks, across which passage had to be made by a plank. As it chanced, this slender bridge, through some accident, had got displaced, one end of the plank being down in the water, so forcing the angler to the alternative of a round-about or wade. That, however, was matter for after-consideration, since what he saw at the moment engrossed all his thoughts—this a heron standing upon the plank, near to the point where it went under the water. With eyes intent on something subaqueous, the bird neither saw nor heard him, as he

HERONS.

had approached with noiseless tread over the soft, grassy turf. In like silence coming to a stop, he took survey of the long-legged bird—*piscator* as himself—continuing to regard it for more than a minute. It might have been longer but for a fancy occurring to him, and yielding to this, he extended his trout rod, and with its tip touched and tickled the heron on the back of the neck. The bird, taken by surprise, seemed absolutely astounded, so much that for some seconds it made no movement, but stood as if dazed and unable to stir from the spot. At length, however, it recovered itself, and, spreading its huge wings, rose up into the air, with a fluttering, eccentric flight and manner so comical that the angler, though alone, could not restrain himself from loud laughter.

WHY WAGE WAR AGAINST THE HAWKS?

No doubt the disciples of æstheticism would back me in the advocacy of protection to our birds of prey— especially the *Falconidæ*. So would any one with a spark of sentiment who has ever watched kite, kestrel, or peregrine winging its way through the "ambient air." The wheelings and spiral windings; the pause on quickly pulsing wings, as if the bird were settled upon a perch, then the rapid downward shoot, as arrow from bow, are all displays of graceful motion,—the very perfection of it,—while the presence of the falcon itself adds an indescribable interest to the scene. Yet, for the sake of a few pheasants or partridges—so few as to be scarce worth consideration—a fellow in a velveteen shooting-coat is empowered to wage constant war upon these beautiful birds—one or more such destroyers in every parish—to the danger of their extirpation and the damage of our scenery !

The whole thing is a stupid mistake, calling for remedial legislation, and loudly too. I am no advocate for the abolition of our Game Laws; quite the contrary. Were they done away with, we would soon have no game to legislate for, and the nation would be the loser thereby, if only in the grosser sense of food produce, to an amount few have any idea of. But there are other tastes to be gratified than that of the palate—other cravings to be considered besides those of the stomach—and, without fearing to be taken for a "too utterly," I venture on saying that, to a man of true refinement and appreciation of the beautiful, the spectacle of one of our *Falconidæ* in flight through upper air were worth more than all the pheasants and partridges it is ever likely to "stoop" down upon.

THE FLIGHT OF BIRDS—HOW TURN THEY SIMULTANEOUSLY ?

Speaking of the flight of falcons leads to thinking about that of other birds; and I am reminded of a large flock of starlings, with another of lapwings, I lately saw close together when out for a drive. I was forcibly struck, though not for the first time, with that peculiarity in the flight of both species, which I believe has never been explained, if indeed ever understood. I mean the whole flock changing course at exactly the same instant, no matter how quick or abrupt the turn, or whether the evolutions be upward, downward, to right or to left,

With wild swans and geese one might suppose the former guided by the whoop or whistle of their leader, and the latter by the well-known "honk," as soldiers by word of command. But no such note seems to direct the movements of either starling, or lapwing, in their wheelings and turnings. Then what does? A question, so far as I am aware, unanswered, if answerable. Will electricity explain it—some biological chain of mind or instinct, binding the birds together, and acting on all simultaneously, or with that rapidity by which the electric fluid runs along the wires?

Whatever be the nature of this singular and unexplained phenomenon, it is not alone confined to birds. Quadrupeds also give illustration of it, as often witnessed in cattle on the American prairies. A herd of a thousand, or more, will be tranquilly browsing—perhaps lying down quietly chewing their cud—when, *presto!* all spring up together, and start off in *stampede*, as if each and all had been stung by gadflies at the self-same instant of time.

Every one who has been to sea must have observed "schools" of fish—herrings or mackerel—act in a similar fashion; while in the insect world we have many examples of the same—notably among ants, and bees at their hiving time. How little do we yet know of nature's workings, even of those that are every day, and in clearest daylight, under our very eyes!

HAWK AND HERON.

To shoot or otherwise kill a Heron should, in my opinion, be made punishable by a fine heavier than any imposed upon poaching. Otherwise this bird will ere long disappear from our islands, as has its beautiful congener, the great white egret. Yet a Heron winging its way through the high heavens, or on a moonlight night standing contemplative by stream or tarn, is a most interesting sight. Alas! one is every day becoming rarer from the bird being popped at by every creature who carries a gun.

In the days of falconry the Heron was accounted noblest of quarry; the species of Hawk usually flown at it being the peregrine falcon—a fine bird also getting fast exterminated by the misdirected zeal of the gamekeeper. Rarely was a single peregrine engaged in the chase, but a pair, or *cast;* as otherwise the would-be victor would often be vanquished. Even when the two assailed it they did not always come off unscathed, the Heron transfixing one or other on its long bayonet-like beak. This would occur when the quarry was brought back to ground; and the first thought of the falconer, after sounding his " whoop! " of triumph, was to whistle off his Hawks, to save them from such impalement. But sometimes, also, in the air has the Heron proved itself the better bird, when the fight was a fair one, and with only a single antagonist. A poetical description of a combat so terminating is appended, with a moral I can recommend :—

SIC SEMPER TYRANNIS.

A Heron flew out of the forest, from the top of a withered pine,
And floated away, like a shadowy cloud, to the west, in a slanting
    line ;

Over the creek, and over the moor, with its drifts of grey lichen
    stone,
Away for the reedy swamp, where he'd oft brooded lorn and
    lone.

A Hawk flew out of the forest, from his perch on a naked bough,
Battling his flight in illuminate air, with no longer a look below,
Dashing in spiral circles the beams as the phosphorent waves of
    the bay,
Till with pencils of light his quivering plumes glittered as star
    in the day.

The Hawk was earl of the forest, and feudal chief of the herne,
No *parvenu*, but a Norman lord; so, when quarrie he did
    discern,
On the rights divine of *Falconidæ* Sir Peregrine took his stand,
And stooped as a lordly emperor stoops on a feeble frontier land.

Wheeling, the Heron, with point to the foe, eye steady, and
    ready stroke,
Watched well and smote, as the flashing Hawk through the
    dazzling sunlight broke,
Struck him inside his carte and tierce, and ere he could parry
    the glance,
Spitted him as a Tartar impaled on a Polish lance.

" Sic semper Tyrannis ! " Thus immutable fate decrees;
Hawk, headlong over and over, falls into the ripple of trees,
While the Heron spreads its pinions, and leisurely crossing the
    creek,
Relights on the branch of the withered pine, and wipes the blood
    from its beak.

INDEX.